Quantum Mechanics: Concepts and Applications

Quantum Mechanics: Concepts and Applications

Edited by
Adaline Cerny

MURPHY & MOORE
www.murphy-moorepublishing.com

Published by Murphy & Moore Publishing,
1 Rockefeller Plaza,
New York City, NY 10020, USA

ISBN: 978-1-63987-470-5

Cataloging-in-Publication Data

Quantum mechanics : concepts and applications / edited by Adaline Cerny.
p. cm.
Includes bibliographical references and index.
ISBN 978-1-63987-470-5
1. Quantum theory. 2. Mechanics. 3. Thermodynamics. I. Cerny, Adaline.
QC174.12 .Q36 2022
530.12--dc23

For information on all Murphy & Moore Publications
visit our website at www.murphy-moorepublishing.com

MURPHY & MOORE

Contents

Preface

The world is advancing at a fast pace like never before. Therefore, the need is to keep up with the latest developments. This book was an idea that came to fruition when the specialists in the area realized the need to coordinate together and document essential themes in the subject. That's when I was requested to be the editor. Editing this book has been an honour as it brings together diverse authors researching on different streams of the field. The book collates essential materials contributed by veterans in the area which can be utilized by students and researchers alike.

Quantum mechanics is a theory in physics that explains the physical properties of nature at the atomic and subatomic levels. It is fundamental to the study of all quantum physics including quantum chemistry, quantum field theory, quantum technology, and quantum information science. Classical physics defines many aspects of nature at an ordinary scale, while quantum mechanics describes the aspects of nature at small scales. Most theories in classical physics can be derived from quantum mechanics as an estimation valid at large scale. Quantum mechanics is used to describe observations which could not be resolved with classical physics. This book attempts to understand the multiple branches that fall under this discipline and how such concepts have practical applications. Different approaches, evaluations, methodologies and advanced studies on quantum mechanics have been included herein. The extensive content of this book provides the readers with a thorough understanding of the subject.

Each chapter is a sole-standing publication that reflects each author's interpretation. Thus, the book displays a multi-facetted picture of our current understanding of application, resources and aspects of the field. I would like to thank the contributors of this book and my family for their endless support.

Editor

1

The Nuclear Mean Field Theory and its Application to Nuclear Physics

M.R. Pahlavani

1. Introduction

Our universe consists of substance. Atoms and molecules are basic components of material. Each atom contains a nucleus which is spread in a small area of atom, and electrons. Also, a nucleus contains Neutrons and protons. It is well known today that electrons in atom and Neutrons and protons in the nucleus are interacting together through different forces. It is clear today that the source of different interactions are composed of four basic forces of the universe, namely gravitational, coulomb, strong and weak nuclear interactions.

In quantum mechanics, to study a particle, it is necessary to have knowledge about its interaction with the surrounding media. The Schrödinger equation is a second-order differential equation that is solved to obtain energy spectrum and wave functions of a particle in quantum mechanics. For a many-body system such as atom or nucleus, it is not possible to solve a set of Schrödinger equations to obtain energy spectrum and wave functions analytically. Therefore in such situations, it is necessary to use an average potential which is a mean potential of all interacting forces acting upon a single particle. Then the Schrödinger equation is should be solve for a single particle. This procedure is called the mean field method [1, 2, 3].

To review this method consider a system consisting of N identical interacting particles. The Hamiltonian of system composed of kinetic energy, T, and potential energy, V, is defined as

$$H = T + V = \sum_{i=1}^{N} t(r_i) + \sum_{\substack{i,j<1 \\ i<j}}^{N} v(r_i, r_j) = \sum_{i=1}^{N} -\frac{\hbar^2}{2m_N}\nabla_i^2 + \sum_{\substack{i,j<1 \\ i<j}}^{N} v(r_i, r_j) \tag{1}$$

where m_N is the mass of each particle, and r_i denotes the coordinates of particle i. A summed single particle potential energy, so far undefined, can be added and subtracted of the Hamiltonian to obtain the following relation,

$$\boldsymbol{H} = \left[\boldsymbol{T} + \sum_{i=1}^{N} \boldsymbol{v}(\boldsymbol{r}_i)\right] + \left[\boldsymbol{V} - \sum_{i=1}^{N} \boldsymbol{v}(\boldsymbol{r}_i)\right] = H_{MF} + \boldsymbol{V}_{RES} \tag{2}$$

where

$$H_{MF} = T + \sum_{i=1}^{N} \boldsymbol{v}(\boldsymbol{r}_i) \equiv T + V_{MF} = \sum_{i=1}^{N} \left[t(r_i) + \boldsymbol{v}(\boldsymbol{r}_i)\right] \equiv \sum_{i=1}^{N} h(r_i) \tag{3}$$

is the mean field Hamiltonian of the system and

$$\boldsymbol{V}_{RES} = V - \sum_{i=1}^{N} \boldsymbol{v}(\boldsymbol{r}_i) = \sum_{\substack{i,j<1 \\ i<j}}^{N} v(r_i, r_j) - \sum_{i=1}^{N} \boldsymbol{v}(\boldsymbol{r}_i), \tag{4}$$

is the mean residual interaction. It should be noted that the residual interaction is related to the strength of the actual interaction and can be reduced if the mean field potential is close to the actual potential of the system.

Actually, the mean field method is an approximation in which each particle of system moves under an external field generated by the remaining $N-1$ particles. This mean potential, V_{MF}, can be considered as an average of all possible interactions of nucleons during the short time interval ΔT, between the selected nucleon and its surrenders,

$$\boldsymbol{V}_{MF} = \sum_{i=1}^{N} \boldsymbol{v}(\boldsymbol{r}_i),\ \boldsymbol{v}(\boldsymbol{r}_i) = \frac{1}{\Delta T} \int_{T}^{T+\Delta T} dt \sum_{\substack{j=i \\ j\neq i}}^{N} v(r_i(t),\ r_j(t)). \tag{5}$$

It is important to know that the time average idea was considered only for clearance of the subject and not applicable in practice unless one studies the thermo-dynamical behavior of nucleus.

Therefore the idea of using mean field theory capable of reducing many particles interacting system in to a system of non-interacting (quasi-particles) considered in an external field, VMF which is the mean potential of possible forces of interaction. The mean field potential is considered such that the stationary Schrödinger equation is solved simply to obtain single particle states and their related energy spectrum. These single-particle states are used to construct the N particle wave function as follows.

The corresponding N -particle Schrödinger equation is used to obtain solutions of the mean-field Hamiltonian H_{MF}

$$H_{MF}\Psi_0(r_1, r_2, \dots r_N) = E\Psi_0(r_1, r_2, \dots r_N). \tag{6}$$

The wave function $\Psi_0(r_1, r_2, \dots r_N)$ can be separated by using the ansatz single particle wave functions

$$\Psi_0(r_1, r_2, \dots r_N) = \phi_{\alpha_1}(r_1)\phi_{\alpha_2}(r_2)\dots\phi_{\alpha_N}(r_N). \tag{7}$$

Substituting this ansatz in to the Schrödinger equation (6) yields N identical one-particle Schrödinger equations

$$\begin{gathered} h(r)\phi_\alpha(r) = \varepsilon_\alpha \phi_\alpha(r), \\ h(r) = t(r) + v(r) = -\frac{\hbar^2}{2m_N}\nabla^2 + v(r). \end{gathered} \tag{8}$$

With the quasi-particle energy, ε_α, that is satisfies the following condition

$$E = \sum_{i=1}^{N} \epsilon_\alpha . \tag{9}$$

The wave function of the many- body system is thus an anti symmetric product of single-particle wave functions which are one-particle wave functions of an external potential well. In summary the mean field theory reduces the complicated many-body problem in to a simple one-particle system.

The main idea in this approach is to determine the mean field potential or in particular, an appropriate mean field potential in which the residual interactions between the quasi-particles should be optimal. To do so, one may seek an optimal set $\{\phi_\alpha(r)\}$ of one-quasi-particle states. This is a Rayleigh-Ritz variational approximation in which the variation $\phi_\alpha(r) \to \phi_\alpha(r) + \delta\phi_\alpha(r)$ of the single-particle orbital is minimized

$$\begin{gathered} E_{gs} = \langle \Psi_0 | H | \Psi_0 \rangle \\ H = T + V_{MF} + V_{RES} \end{gathered} \tag{10}$$

As a starting point, one may construct an ansatz wave function. It is customary to use a product of single particle wave functions as Eigen function of the system,

$$\Psi_0(r_1, r_2, \ldots r_N) = \prod_{i=1}^{N} \phi_{\alpha_i}(r_i) \tag{11}$$

It is an anti-symmetrized product ansatz wave function following the Hartree-Fock method and is called the Slater determinant of the given single particle states

$$\Psi_0(r_1, r_2, \ldots r_N) = C\left[\prod_{i=1}^{N} \phi_{\alpha_i}(r_i)\right]. \tag{12}$$

Here $\Psi_0(r_1, r_2, \ldots r_N)$ is an anti-symmetric wave function. Also C is the normalization constant. For instance, consider a three-particles system with its single-particle Eigen states labeled 1, 2, and 3. Then the normalized anti-symmetric state, or the Slater determinant, is

$$\Psi_0(r_1, r_2, \ldots r_N) = \frac{1}{\sqrt{6}} \begin{vmatrix} \phi_{\alpha_1}(r_1) & \phi_{\alpha_1}(r_2) & \phi_{\alpha_1}(r_3) \\ \phi_{\alpha_2}(r_1) & \phi_{\alpha_2}(r_2) & \phi_{\alpha_2}(r_3) \\ \phi_{\alpha_3}(r_1) & \phi_{\alpha_3}(r_2) & \phi_{\alpha_3}(r_3) \end{vmatrix}. \tag{13}$$

The energy E of the system has to be varied under the constraint that the normalization of Ψ_0 is preserved, i.e. $\langle\Psi_0|\Psi_0\rangle = 1$. This leads to the constrained variational problem,

$$\delta\left(\frac{\langle\Psi_0|H|\Psi_0\rangle}{\Psi_0|\Psi_0}\right) = 0,$$

which can be transformed in to an unconstrained one by minimizing the energy for normalized wave function, $\Psi_0(r_1, r_2, \ldots r_N)$. After performing the variation, the single-particle energy, ε_α, is can also be obtained.

One powerful method to address such uncertainties is the following Hartree consistent equation [4,5],

$$-\frac{\hbar^2}{2m_N}\nabla^2\phi_\alpha(r) + V_{H(F)}(\{\phi_i\})\phi_\alpha(r) = \epsilon_\alpha\phi_\alpha(r),$$
$$i = 1, 2, \ldots\ldots N, \; \alpha = 1, 2, \ldots\ldots\infty \tag{14}$$

This equation is like the Schrödinger equation except that the simple potential term, V (r), is replaced with a function of unknown wave function,

$$V(r) = V_{H(F)}\left(\{\phi_i(r)\}\right)$$

Here, the Hartree mean field potential, $V_{H(F)}$, is different from, V_{HF}, Hartree-Fock mean field.

The differential equation (14) is nonlinear and therefore, much more difficult to solve than the regular Schrödinger equation. The solution can only be carried using consistent iteration method. In this procedure, one can start using a complete set of guessed single-particle states $\{\phi_i^0\ (r)\}$, $i=1, \ldots.N$ to calculate the initial potential term, $V_{H(F)}^{(0)}$. In the next step, the equation for a complete set of new wave functions $\{\phi_\alpha^{(1)}\ (r)\}$ $\alpha=1, \ldots..\infty$ is solved to obtain Eigen energies $\varepsilon_\alpha^{(1)}$. The procedure is then repeated with new Eigen function $\phi_\alpha^{(1)}\ (r)$ to obtain the new potential , $V_{H(F)}^{(1)}$. This approach can be depicted through the following schematic diagram,

$$\begin{gathered} \phi_i^0(r) \rightarrow V_{H(F)}^{(0)} \rightarrow \epsilon_\alpha^{(1)}, \\ \epsilon_\alpha^{(1)} \xrightarrow{\phi_i^{(1)}} V_{H(F)}^{(1)} \rightarrow \ldots \rightarrow \phi_\alpha^{(n)},\ \epsilon_\alpha^{(n)}. \end{gathered} \tag{15}$$

This procedure is repeated to achieve self-consistency for wave functions (or Eigen energies). This means that after each loop the resultant wave function or Eigen energies compared with the starting wave function or Eigen energies and when their difference becomes less than a given preset limit, i.e.

$$\left\|\phi_\alpha^{(n-1)} - \phi_\alpha^{(n)}\right\| < preset\ limit,$$

the procedure is repeated, otherwise, it will be automatically terminated. Where the $\| \ldots \|$ denotes the norm.

The results of each run, contain a self-consistent mean field, $V_{H(F)}(r)$, the Eigen state, $\phi_\alpha(r)$, and its associated Eigen energies, ε_α, are all simultaneously generated. We may also note that for a finite potential-well, there will be, in addition to the bound states, an infinite number of unbound states.

In our discussions, the generated mean-field potential is a central one, that is only a function of r. Central mean field potentials describe only systems with spherical symmetry such as spherical nuclei or atoms. This is because of natural real forces that are conservative and satisfy the conservation of energy.

In some convenient way to avoid self-consistency loops, a phenomenological potential like a simple square well with finite depth, simple harmonic oscillator well and complicated Woods-Saxon with considerable parameters that can be determined using the fit of potential with experimental data, is introduced.

2. Applications of mean-field theory in nuclear physics

Over the years after Rutherford's valuable experiments that suggest nuclei for atom, many theoretical and experimental attempts have been done to obtain knowledge about the stability of nuclei. It is clear today that a nucleus of mass number A, Neutron number N and proton number (atomic number) Z, consists of A strongly interacting nucleons (protons and neutrons in the nucleus without considering their different properties called nucleon.). In addition to the strong nuclear force that is responsible for nuclear stability, the protons also sense the attractive coulomb potential because of their charge. In regular nuclear physics, the protons and neutrons are considered the point particles without any internal structure. This is an excellent approximation when the aim is to study nuclear structure at low energies. In such approach, the nuclear forces are considered a central attractive force with proper specifications like independence of charge and low range. Note that in advance models of nuclear physics such as the Yukawa Meson exchange model, it is believed that nucleons constructed quarks and interact together through the meson exchange mechanism in the base of the particle physics lows. The lightest nucleus is Deuterium with one neutron and one proton. The interaction of nucleons in the nucleus can be studied both theoretically and practically using simple Deuteron nucleus. This two-nucleon system is described by two-body interaction matrix elements, without a detailed account of the methods used to obtain them. On the other hand, the A - nucleons nucleus in quantum mechanics using the Schrödinger equation is not a solvable problem analytically at least for $A>10$. Therefore, one has to look for a reasonable approximate method to solve this many-body problem consisting of strongly interacting nucleons. A powerful approximation is to convert such many-body system in to a non-interacting system of quasi-particles using a suitable external mean field potential. The remaining interactions, called residual interaction, can be treated as a perturbation potential in the base of perturbation approximation. As discussed earlier, the transformation of system of particles in to quasi-particles is not simple, and its success depends on the nuclear system under consideration.

As mentioned above, a conventional approach is to select a particular type of mean field potential to avoid the steps leading to self-consistency. The selected mean field potential and considered remaining residual interactions as approximations produce the preciseness of the obtained results. The simplest custom potential is the three-dimensional harmonic oscillator potential well

$$\mathrm{V}_{HO}(\mathrm{r}) = -\mathrm{V}_1 + \mathrm{kr}^2 = -\mathrm{V}_1 + \frac{1}{2}m\omega^2 r^2 \tag{16}$$

where V_1 and k are the parameters to be fitted to the practical data for best result. A common, more realistic choice is the Woods–Saxon potential [6]

$$v_{WS}(r) = \frac{-V_0}{1+e^{(r-R)/a}}.$$

where V_0, R and a are the nuclear potential depth, the nuclear radius, and the surface diffuseness, respectively. They are parameterized as follows,

$$V_0=\left(51\pm 33\frac{N-Z}{A}\right)MeV,\ R=r_0A^{\frac{1}{3}}=1.27A^{\frac{1}{3}}\ fm,\ a=0.67\ fm.$$

The + and – signs are considered for protons and neutrons, respectively. In the case when there is no distinction between protons and neutrons a suitable average value of $V_0 = 57\,MeV$ can be used for nucleons.

The Woods–Saxon potential, v_{WS}, is a suitable choice for the mean field potential however it is a complicated function of, r, and it is not an analytically solvable one. To overcome this problem, it is possible to select the proper three- dimensional oscillator potential with energy quantum, $\hbar\omega$ and depth, V_1. The energy difference of levels, $\hbar\omega$, and depth, V_1, can be obtained with a best fit to the Woods–Saxon potential, v_{WS}, as a function of, V_0, R and a as the nuclear potential depth, nuclear radius, and surface diffuseness of the Woods–Saxon potential, respectively. The wave functions and energy spectrum of equivalent harmonic oscillator potential agreed well with the Woods–Saxon potential ones especially near the bottom of the wells in low energies. The difference of these potentials increases when the potential approaches zero. Actually the major difference of these potentials is that the harmonic oscillator potential varies more sharply than the Woods–Saxon one near the surface of the nucleus.

2.1. The spin–orbit interaction

Sometimes in 1949, Meyer and independently, Haxel, Jensen, and Swees showed that if in addition to mean field central potential, V_{MF}, a non-central potential is included in the Schrödinger equation, all closed shell nucleon numbers can be obtained successfully. These numbers 2, 8, 20, 28, 50, 82, and 126 are called magic numbers because the origin of these numbers was not known at that time. The Woods-Saxon or its equivalent harmonic oscillator central potential is not able to reproduce experimentally observed precise data of the single-particle structure energies of the nucleus using the mean field approach.

The non-central potential due to the interaction between the spin of nucleons with the angular momentum of orbital that nucleons located on it, is called spin-orbit interaction. As a result of spin-orbit interaction [7, 8], the nuclear energy level for a given l (except for $l=0$) is split in to two sublevels. The sublevels are characterized by total angular momentum numbers equal to $\left(l+\frac{1}{2}\right)$ and $\left(l-\frac{1}{2}\right)$ corresponding to whether the spin is parallel or anti-parallel to the orbital angular momentum. Each sublevel with spin j accommodates $(2j+1)$ neutrons or protons. The same interaction with a different structure is observed in atoms with a different sign as in the nucleus.

Consider that the harmonic oscillator central potential is produced only for the first three observed magic numbers 2, 8, and 20. To obtain the remaining numbers 28, 50, 82 and 126, it is necessary to add a spin-orbit interaction potential to the Schrödinger equation.

The origin of the spin–orbit interaction is not the same in atoms and nucleus. The atomic spin–orbit force is due to a well-known electromagnetic interaction, and the scale of energy separation is in the order of milli-electronvolts, while the energy difference of sublevels separated because of the nuclear spin-orbit interaction is in the order of million electronvolts and its origin is not well understood yet. In most cases, this force is considered phenomenologically. For the spin–orbit term, we use [9]

$$v_{LS}(r)=v_{LS}^{0}\left(\frac{r_0}{\hbar}\right)^2\frac{1}{r}\frac{d}{dr}\left[\frac{1}{1+e^{(r-R)/a}}\right] \tag{17}$$

The second pair of parentheses guarantees that the derivative does not operate on the wave function when substituted in the radial Schrödinger equation. The r dependence of this interaction arises from its central nature.

The derivative part of this potential is often neglected for simplicity and $v_{LS}(r)$ is replaced by a constant; however, to obtain precise results, the radial part should be considered. We have

$$v_{LS}^{0} = 0.44V0.$$

To obtain the strength of the spin-orbit part, we use

$$J^2=(L+S)^2$$
$$J^2=L^2+S^2+2L.S,$$

and its expectation value for the nuclear wave equation made

$$<L.S>=\frac{\hbar^2}{2}\left[j(j+1)-l(l+1)-\frac{1}{2}(\frac{1}{2}+1)\right]$$
$$<L.S>=\frac{\hbar^2}{2}l \quad for \quad j=l+\frac{1}{2},$$
$$<L.S>=\frac{\hbar^2}{2}(-l-1) \quad for \quad j=l-\frac{1}{2}.$$

In addition to the mean field plus spin–orbit interaction, protons in nuclei interact together via the coulomb force, which is defined by the following relation, considering nuclei as a sphere with a constant charge density [10]

$$V_C(r)=\frac{Ze^2}{4\pi\varepsilon_0}\begin{cases}\dfrac{3-\left(\dfrac{r}{R}\right)^2}{2R} & r\le R\ , \quad \dfrac{1}{r} \quad r\ge R\end{cases} \tag{18}$$

To obtain the energy spectrum and wave functions for neutrons, one needs to solve the radial Schrödinger equation for the Woods-Saxon and spin-orbit potentials. Such second-order differential equation cannot be solved analytically. To solve this complicated differential equation, it is necessary to introduce some new variables and use reasonable approximations. By introducing new variable [11] $y=\frac{1}{1+\exp\frac{(r-R)}{a}}$, the Woods–Saxon potential reduces to its simple form $V_{WS}=V_{0y}$ while the spin-orbit term changes to $V_{LS}=\frac{(y-y^2)}{R_0+a\ln\left(\frac{1}{y}-1\right)}$. For orbits with small l, the Taylor expansion of the $\frac{1}{r}$ near $r=r_m$, is reasonable. According to the definition of variable y we have, $f(y)\equiv\frac{1}{1+\exp\frac{(r-R)}{a}}$ hence by expanding $f(y)$ around $y_m\equiv\frac{1}{1+\exp\frac{(r_m-R_0)}{a}}$ with $0<y_m<1$, since $0<y<1$, y^3 and the higher terms are negligible, the radial part of the spin-orbit term can be approximated using [12],

$$-\frac{a}{r}\frac{d}{dr}\left(\frac{1}{1+\exp\frac{(r-R)}{a}}\right)=\frac{1}{r_m}\left(C_0+C_1y+C_2y^2\right), \tag{19}$$

where C_0, C_1, and C_2 are dimensionless coefficients and evaluated as

$$C_0=\frac{-ar_my_m+2a^2y_m}{2r_m^2(1-y_m)},$$
$$C_1=1+\frac{ar_my_m+2a^2}{r_m^2(1-y_m)},$$
$$C_2=-1+\frac{ar_m(1+2y_m)+2a^2}{r_m^2(1-y_m)2y_m}.$$

Likewise, the Taylor expansion is applicable for $\frac{1}{r^2}$,

$$\frac{1}{r^2}\approx\frac{1}{r_m^2}\left(D_0+D_1y+D_2y^2\right),$$

where D_i's, similar to the C_i', are obtained through

$$D_0=1-\frac{4a}{r_m(1-y_m)}+\frac{3a^2+ar_m}{r_m^2(1-y_m)^2},$$
$$D_1=\frac{6a}{r_my_m(1-y_m)}-\frac{6a^2+2ar_m}{r_m^2y_m(1-y_m)^2},$$
$$D_2=\frac{3a^2-ar_m(1-2y_m)}{r_m^2(1-y_m)2y_m^2}.$$

This type of expansion has been widely used for differential equations resulting from the Schrödinger equation with different potentials [12].

By means of these expansions, the spin-orbit term transforms into

$$\text{VLS} = (\text{C0} + \text{C1y} + \text{C2y2}),$$

and the centrifugal term is obtained

$$\text{Vc.f=(D0+D1y+D2y2).}$$

By using these expansions, the spin-orbit term transforms into $V_{LS} \propto (C_o + C_1 y + C_2 y^2)$ and the centrifugal term is changed to the favorable type $V_{CF} \propto (D_o + D_1 y + D_2 y^2)$. The substitution of V_{LS} and V_{CF} as a function of variable y into the Schrödinger equation transforms this equation in to the following analytically solvable differential equation

$$\begin{aligned} &y(1-y)\left[y(1-y)\frac{d^2R(y)}{d\,y^2} + (1-2y)\frac{dR(y)}{d\,y^2}\right] + \\ &+\frac{2ma^2}{\hbar^2}[V_{0y} + V_{LS}^{'(0)}\left(C_o + C_1 y + C_2 y^2\right) \\ &-\frac{\hbar^2}{2m}\frac{l(l+1)}{r_m^2}\left(D_o + D_1 y + D_2 y^2\right) R(y) = 0 \\ &V_{LS}^{'(0)} = \frac{r_0^2 V_{LS}^{(0)}}{2ar_m}(j(j+1) - l(l+1) - 3/4) \end{aligned}$$

This equation can be transformed into the following simple form,

$$y(1-y)\frac{d^2R(y)}{dy^2} + (1-2y)\frac{dR(y)}{dy} + \frac{-\varepsilon^2 + \beta^2 y - \gamma^2 y^2}{y(1-y)}R(y) = 0 \tag{20}$$

Equation (20) can be transformed into the well-known form of hypergeometric differential equation or, alternatively Nikiforo-Avorono (NU) type [13]. The obtained results using the NU method are

$$-\varepsilon^2 = \frac{2ma^2}{\hbar^2}\left(E + V_{LS}^{'(0)}\right) - \frac{l(l+1)}{r_m^2} D_0$$

$$\beta^2 = \frac{2ma^2}{\hbar^2}\left(V_0 + V_{LS}^{'(0)}C_1\right) - \frac{l(l+1)}{r_m^2} D_1$$

$$\gamma^2 = -\frac{2ma^2}{\hbar^2}\left(E + V_{LS}^{'(0)}C_2\right) - \frac{l(l+1)}{r_m^2} D_2$$

$$R = C\frac{\Gamma(2\varepsilon-1)\Gamma(-2i\lambda)}{\Gamma(\varepsilon+\eta'+1-i\lambda)\Gamma(\varepsilon-\eta'-i\lambda)}\left[(1-y)^{i\lambda} + \frac{\Gamma(2i\lambda)\Gamma(\varepsilon-\eta'+1-i\lambda)\Gamma(\varepsilon-\eta'-i\lambda)}{\Gamma(-2i\lambda)\Gamma(\varepsilon-\eta'+1+i\lambda)\Gamma(\varepsilon-\eta'+i\lambda)}(1-y)^{-i\lambda}\right]$$

where Γ is the well-known gamma function, and C is the normalization constant. λ, μ, and η' are defined as follows

$$\lambda = \sqrt{\beta^2 + \varepsilon^2 + \gamma^2}$$

$$\mu = i\lambda$$

$$\eta' = \sqrt{\gamma^2 - 1/4} - 1/2$$

Note that λ is valid only for the $\beta^2 > \varepsilon^2 + \gamma^2$ condition. In a special case where *l=0,* the solution reduces to its simple form. Also, the energy eigenvalues are obtained as a function of z satisfying the following relation

$$-\frac{z}{\sqrt{1-z^2}} = \tan\varnothing(z),\ z = \sqrt{1-\frac{|E|}{V_0}}$$

which *Φ* (z) can be evaluated using a graphical method.

$$\varnothing(z) = k_0R_0z + \\ +\tan^{-1}\left(\frac{z}{\sqrt{1-z^2+\left(\frac{1}{k_0a}\right)}}\right) - \sum_{n=0}^{\infty}\left[\tan^{-1}\left(\frac{2k_0a}{n}z\right) - \tan^{-1}\left(\frac{z}{\sqrt{1-z^2}+\left(\frac{n}{k_0a}\right)}\right) - \tan^{-1}\left(\frac{z}{\sqrt{1-z^2}+\left(\frac{n+1}{k_0a}\right)}\right)\right].$$

Note that

$$k = \frac{\lambda}{a} = \left[\frac{2m}{\hbar^2}(V_0' - |E|\right]^{1/2},$$

and

$$k_0 = \frac{2m}{\hbar^2}(V_0^{'})^{1/2}\ ,\ z = \frac{k}{k_0}.$$

Finally,

$$|E| = V_0\left(1 - Z^2\right)$$

The results obtained in this special case are in agreement with the results obtained using other methods [14].

3. Conclusions

In this chapter we briefly discussed the idea of mean field theory as an improvable approximation method for many-body problems of identical particles like atoms and nucleus that cannot be solved analytically. We have shown that for a system of A - nucleons nucleus by considering a suitable potential using this model, one is able to obtain energy spectrum and wave equations. However, the obtained results cannot reproduce the measured nuclear spectroscopy, but one may hope to become successful by considering an accurate potential in the Schrödinger equation.

Author details

M.R. Pahlavani

Address all correspondence to: m.pahlavani@umz.ac.ir

Department of Nuclear Physics, Faculty of Basic Science, University of Mazandaran, Babolsar, Iran

References

[1] Suhonen Jouni. From nucleon to nucleus. Springer-Verlag Berlin Heidelberg (2007).

[2] Ring P, Schuck P. The nuclear many-body problem. Springer-Verlag New York Heidelberg Berlin, pages 126, 314 and 438 (1980).

[3] Kelban M, et al., Global properties of spherical nuclei obtained from Hartree-Fock-Bogoliubov calculations with the Gogny force, Phys. Rev. C 65 (2002) 024309.

[4] Hartree D R, Hartree-Fock theory, Proc. Cambridge Phil. Soc. 24 (1928) 89.

[5] Slater J C, A simplification of Hartree-Fock method, Phys. Rev. 81 (1951) 385.

[6] Judek J, et al., Parameters of the deformed Wioods-Saxon potential outside A=110-210 nuclei, Journal of physics G: Nuclear and Particle physics, 5 (1979) 1359.

[7] Asmart Mahmoud and Ulloa Sergio, Spin-orbit interaction and isotropic electronic transport Graphen, Phys. Rev. Let. 112 (2014) 136602.

[8] Glazov M M, Sherman E ya and Dugaev V K, Two-dimensional electron gas with spin orbit disorder, Physica E 42 (2010) 2157.

[9] Hull Jr M H, Dyatt Jr K D, Fisher C R and Brelt G, Nucleon-Nucleon spin-orbit interaction potential, Phys. Rev. Let. 2 (1959) 264.

[10] Eder G and Berhummer O, The coulomb potential of spherical nuclei, Letters Al. Nuovo Cimento 15 (1976) 25.

[11] Pahlavani M R and Alavi S A, Solutions of Woods-Saxon potential with Spin-Orbit and centrifugal terms through Nikiforo-Uvarov method, Commun. Theor. Phys. 588 (2012) 793.

[12] Pahlavani M R and Alavi S A, Study of nuclear bound states using mean-field Woods-Saxon and spin orbit potentials, Modern Physics Letter A 27 (2012) 1250167.

[13] Cüneyt Berkdemir, Application of the Nikiforov-Uvarov Method in Quantum Mechanics, Theoretical Concepts of Quantum Mechanics, Prof. Mohammad Reza Pahlavani (Ed.), ISBN: 978-953-51-0088-1, InTech, (2012). DOI: 10.5772/33510.

[14] Lu G, analytic quantum mechanics of diatomic molecules with empirical potentials, Physics Scripta, 72 (2005) 349.

2

Implications of Quantum Informational Entropy in Some Fundamental Physical and Biophysical Models

Maricel Agop, Alina Gavriluţ, Călin Buzea,
Lăcrămioara Ochiuz, Dan Tesloianu,
Gabriel Crumpei and Cristina Popa

1. Introduction

Complex systems are a large multidisciplinary research theme that has been studied using a combination of fundamental theory, derived especially from physics and computational modeling. This kind of systems is composed of a large number of elemental units that interact with each other, being called "agents" [1, 2, 62]. Examples of complex systems can be found in human societies, the brain, internet, ecosystems, biological evolution, stock markets, economies and many others.

The manner in which such a system manifests can't be predicted only by the behavior of individual elements or by adding their behavior, but is determined by the way the elements interact in order to influence global behavior. Very important properties of complex systems are those of emergence, self-organization, adaptability etc. [3, 4, 62].

An example of a complex system is represented by polymers. [Their structures present a multitude of organized networks starting from simple, linear chains of identical structural units to very complex sequences of amino acids that are chained together, thus forming the fundamental units of living fields. Probably one of the most interesting biological complex system is DNA that generates cells by employing a simple but very efficient code. It is the striking way in which individual cells organize into complex systems, such as organs and, subsequently, organisms. Research in the field of complex systems could provide new information on the realistic dynamics of polymers, solving troublesome problems such as protein folding. We note that the dynamics of such complex systems implies the quantum formalism] [1-4, 62].

Correspondingly, the theoretical models that describe the complex systems dynamics become more and more advanced [1-4]. For all that, this problem can be solved by taking into account that the complexity of the interaction process implies various temporal resolution scales, and the pattern evolution implies different degrees of freedom [5].

[In order to develop new theoretical models we must state the fact that the complex systems displaying chaotic behavior are recognized to acquire self-similarity (space-time structures can appear) in association with strong fluctuations at all possible space-time scales [1-4]. Afterwards, for temporal scales that are large with respect to the inverse of the highest Lyapunov exponent, the deterministic trajectories are replaced by a set of potential trajectories and the concept of definite positions by that of probability density] [62]. An interesting example is the collisions processes in complex systems, where the dynamics of the particles can be described by non-differentiable curves.

Since non-differentiability can be considered a universal property of complex systems, it is mandatory to develop a non-differentiable physics. In this way, by considering that the complexity of the interaction processes is replaced by non-differentiability, using the entire range of quantities from the standard physics (differentiable physics) is no longer required [19].

This topic was developed in the Scale Relativity Theory (SRT) [6, 7] and in the non-standard Scale Relativity Theory (NSSRT) [8-22]. [In the framework of SRT or NSSRT we assume that the movements of complex system entities take place on continuous but non-differentiable curves (fractal curves) so that all physical phenomena involved in the dynamics depend not only on the space-time coordinates but also on the space-time scales resolution. In this conjecture, the physical quantities that describe the dynamics of complex systems can be considered as fractal functions. In addition, the entities of the complex system may be reduced to and identified with their own trajectories. In this way, the complex system's behavior will be identical to the one of a special interaction-less "fluid" by means of its geodesics in a non-differentiable (fractal) space] [6, 7, 62].

In such context notions as informational entropy, Onicescu informational energy etc become important in the Nature description. These notions will be correlated with the fractal part of the physical quantities that describe the dynamics of complex systems.

2. Informational entropy and energy

Independently of scale resolution, the motion, either on infragalactic scale (for instance, the planetary motion), or on atomic scale (for instance, the motion of the electron around its nucleus) takes place on conics (ellipses). Such motion in invariant with respect to the SL(2R) group. In what follows, we shall consider this invariance only with respect to the motion on atomic scale.

2.1. SL(2R) invariance and canonic formalism

SL(2R) group is the group of transformations [23-26]

$$x' = \alpha x + \beta y, \quad y' = \gamma x + \delta y, \quad \alpha\delta - \beta\gamma = 1 \tag{1}$$

which makes invariant the areas in the phase space (x, y).

Choosing

$$\alpha = 1 + \frac{1}{2}a_2, \quad \beta = a_1, \quad \gamma = -a_3, \quad \delta = 1 - \frac{1}{2}a_2, \tag{2}$$

the infinitesimal transformations of the group have the expressions

$$x' = x + ya_1 + \frac{p}{2}a_2, \quad y' = y - \frac{1}{2}ya_2 - xa_3 \tag{3}$$

Then the Lie algebra associated to the group becomes

$$[\hat{L}_1, \hat{L}_2] = \hat{L}_1; \quad [\hat{L}_2, \hat{L}_3] = \hat{L}_3; \quad [\hat{L}_3, \hat{L}_1] = -2\hat{L}_2 \tag{4}$$

where

$$\hat{L}_1 = y\frac{\partial}{\partial x}, \quad \hat{L}_2 = \left(x\frac{\partial}{\partial x} - y\frac{\partial}{\partial y}\right), \quad \hat{L}_3 = -x\frac{\partial}{\partial y} \tag{5}$$

are the vectors of the Lie base.

The general vector of the algebra (4) is given by the linear combination

$$\hat{L} = c\hat{L}_1 + 2b\hat{L}_2 + a\hat{L}_3, \quad a, b, c = \text{const.} \tag{6}$$

The hamiltonian H results as an invariant function along the tangent trajectories to the vector (6). Precisely, it is a solution of the equation

$$\hat{L}H = 0 \tag{7}$$

According to (5), relation (7) becomes

$$(bx + cy)\frac{\partial H}{\partial x} - (ax + by)\frac{\partial H}{\partial y} = 0 \tag{8}$$

whence the characteristic differential system

$$\frac{dx}{bx+cy} = -\frac{dy}{(ax+by)} = dt \tag{9}$$

admits the integral

$$H(x,y) = \frac{1}{2}\left(ax^2 + 2bxy + cy^2\right) \tag{10}$$

We notice that the differential system (9) is Hamilton's system of equations [25]

$$\dot{x} = \frac{\partial H}{\partial y}, \quad \dot{y} = -\frac{\partial H}{\partial x} \tag{11}$$

associated to the hamiltonian (1), where the symbol " $\cdot$ " refers to the derivative with respect to the time.

Among the solutions of the equation (1), we have also the Gaussian

$$\rho(x,y) = A\exp[-H(x,y)], \quad A = \text{const.} \tag{12}$$

In consequence, all invariant functions on the group (7) will be functions of the hamiltonian (10) and particularly, of the Gaussian (12).

If the quadratic form (10) is positive definite, that is, the condition

$$a > 0, \ \ \Omega = ac - b^2 > 0 \tag{13}$$

is fulfiled, then, by deriving the relations

$$\begin{aligned} \dot{x} &= bx + cy \\ \dot{y} &= -(ax+by) \end{aligned} \tag{14}$$

with respect to the time and eliminating $\dot{p}$ and $\dot{q}$, based on the relations (14), we obtain the symmetric equations

$$\begin{aligned} \ddot{x} + \Omega x &= 0 \\ \ddot{y} + \Omega y &= 0 \end{aligned} \tag{15}$$

These equations are formally equivalent to the equations of two linear oscillators of coordinates x, y.

Then the 2-form

$$\omega = dx \Lambda dy \tag{16}$$

has the meaning of the elementary surface in phase space (x, y) and the transformations (1) are canonic because they maintain the 2-form (16) (Liouville's theorem [25]). Simultaneously, the Gaussian (12) can be considered as a probabilistic density in phase space (x, y). In this situation, the parameters (a, b, c) can get statistical significance (see also [71]).

2.2. Shannon's informational entropy and transitivity manifolds

[In standard quantum mechanics, the impossibility of determining the variances of the position coordinate Δy_i and of the conjugate momentum component Δx_i ($x_i = -i\hbar\nabla_i$) with arbitrary accuracy is widely accepted as being caused by the unavailable perturbation exerted on the particle by the measuring process. Because the measuring apparatus is most often not defined quantitatively and its perturbation can be very large, the uncertainty relation is formulated as a larger-than-or equal to equation

$$\Delta x_i \Delta y_i \geq \frac{1}{2}\hbar$$

Relating to this, it is unusual that the definitive nonzero variances Δx_i and Δp_i can be obtained for quantum system which are not exposed to a measuring device. This has been shown using the so called negative-result experiments. Furthermore, it can be noticed that we could theoretically obtain the nonzero variance Δx_i and Δy_i of quantum systems without including in the analysis perturbations from or in presence of a measuring device at all] [65].

In such a conjecture the uncertainty relations result in a quite natural way from the momentum perturbations associated with the fractal potential, i.e. with the Shannon's information.

Indeed, let be the probability density in the phase space, $\rho(x, y)$ with the constraints [27-32, 62]

$$\begin{aligned}
&\iint y\rho(x,y)dxdy = \langle y \rangle \\
&\iint x\rho(x,y)dxdy = \langle x \rangle \\
&\iint (y - \langle y \rangle)\rho(x,y)dxdy = (\delta y)^2 \\
&\iint (x - \langle x \rangle)\rho(x,y)dxdy = (\delta x)^2 \\
&\iint (y - \langle y \rangle)(x - \langle x \rangle)\rho(x,y)dxdy = cov(x,y)
\end{aligned} \tag{17}$$

where $\langle y\rangle$ is the mean value of the position, $\langle x\rangle$ is the mean value of the momentum, δy is the position standard deviation, δx is the momentum standard deviation and $cov(x, y)$ is the covariance of the random variables (x, y)] [62].

Now, we introduce Shannon's informational entropy [27]:

$$\bar{H} = \iint \rho(x,y)\ln[\rho(x,y)]dxdy. \tag{18}$$

Through Shannon's maximum informational entropy principle

$$\delta\bar{H} = 0 \tag{19}$$

with constraints (17), we get the normalized Gaussian distribution:

$$\rho\big(x-\langle x\rangle, y-\langle y\rangle\big) = \frac{\sqrt{ac-b^2}}{2\pi}\exp[-H\big(x-\langle x\rangle, y-\langle y\rangle\big)] \tag{20}$$

with

$$\begin{aligned} &H\big(x-\langle x\rangle, y-\langle y\rangle\big) = \frac{1}{2}[\bar{a}(x-\langle x\rangle)^2 + \\ &\quad + 2\bar{b}(x-\langle x\rangle)(y-\langle y\rangle) + \bar{c}(y-\langle y\rangle)^2] \\ &\bar{a} = \frac{(\delta y)^2}{D},\ \bar{b} = -\frac{cov(x,y)}{D},\ \bar{c} = \frac{(\delta x)^2}{D} \\ &D = (\delta x)^2(\delta y)^2 - cov^2(x,y) \end{aligned} \tag{21}$$

[We must note that the set of parameters $(\bar{a}, \bar{b}, \bar{c})$ has statistical significance given by relations (21).

In such context, the statistical hypothesis are specified through a particular choice of the set of parameters $(\bar{a}, \bar{b}, \bar{c})$ of the quadratic form the first Eq (21). Their class is given by the restriction [70]:

$$H(x',y') = H(x'-\langle x\rangle, y'-\langle y\rangle) = H(x-\langle x\rangle, y-\langle y\rangle) \tag{22}$$

where

$$\begin{aligned} &H(x'-\langle x\rangle, y'-\langle y\rangle) = \\ &= \frac{1}{2}\Big[\bar{a}'(x'-\langle x\rangle)^2 + 2\bar{b}'(x'-\langle x\rangle)(y'-\langle y\rangle) + \bar{c}'(y'-\langle y\rangle)^2\Big] \end{aligned} \tag{23}$$

If $(x'-\langle x\rangle,\ y'-\langle y\rangle)$ and $(x-\langle x\rangle,\ y-\langle y\rangle)$ are dependent through the unimodular transformations (1), we get that (22) imposes through $(\bar{a},\ \bar{b},\ \bar{c})$, the group of three parameters] (see [71] for details)

$$\begin{aligned} \bar{a}' &= \delta^2\bar{a} - 2\gamma\delta\bar{b} + \gamma^2\bar{c} \\ \bar{b}' &= -\beta\delta\bar{a} - (\beta\gamma + \alpha\delta)\bar{b} - \alpha\gamma\bar{c} \\ \bar{c}' &= \beta^2\bar{a} - 2\alpha\beta\bar{b} + \alpha^2\bar{c} \end{aligned} \tag{24}$$

If for the group (24) we choose the same parameterization as the one given by relations (2), then the corresponding infinitesimal transformations

$$\begin{aligned} \bar{a}' &= \bar{a} - \bar{a}a_2 + 2\bar{b}a_3 \\ \bar{b}' &= \bar{b} - \bar{a}a_1 + \bar{c}a_3 \\ \bar{c}' &= \bar{c} - 2\bar{b}a_1 + \bar{c}a_2 \end{aligned} \tag{25}$$

can be considered as an incompatible algebraic system in the unknowns $a_1,\ a_2,\ a_3$. In consequence, there cannot exist a transformation able to ensure the correspondence

$$\left(\bar{a}', \bar{b}', \bar{c}'\right) \to \left(\bar{a}, \bar{b}, \bar{c}\right) \tag{26}$$

Thus, the action of the group (24) in the space of variables $(\bar{a},\ \bar{b},\ \bar{c})$ is intransitive and, therefore, there exists a relation among the parameters $(\bar{a},\ \bar{b},\ \bar{c})$, which remains invariant to the action of the group (24). This relation is called transitivity manifold [26] (see also [71]).

The Lie algebra associated to the group (24) is

$$[\hat{A}_1, \hat{A}_2] = \hat{A}_1;\ [\hat{A}_2, \hat{A}_3] = \hat{A}_3;\ [\hat{A}_3, \hat{A}_1] = -2\hat{A}_1 \tag{27}$$

where

$$\begin{aligned} \hat{A}_1 &= -\bar{a}\frac{\partial}{\partial\bar{b}} - 2\bar{b}\frac{\partial}{\partial\bar{c}} \\ \hat{A}_2 &= -\bar{a}\frac{\partial}{\partial\bar{a}} + \bar{c}\frac{\partial}{\partial\bar{c}} \\ \hat{A}_3 &= 2\bar{b}\frac{\partial}{\partial\bar{a}} + \bar{c}\frac{\partial}{\partial\bar{b}} \end{aligned} \tag{28}$$

are the vectors of the base Lie. By the conditions

$$\hat{A}_1 F = 0, \quad \hat{A}_2 F = 0, \quad \hat{A}_3 F = 0 \tag{29}$$

[where F is an arbitrary function, we can obtain the transitivity manifolds of the group in the form

$$\bar{a}\,\bar{c} - \bar{b}^2 = \text{const.} \tag{30}$$

If H has energy significance, then condition (30) shows that a representative point from space $(x,\ y)$ (which is in motion on a surface of constant energy (22)), can be also found on a surface of constant probabilistic density (ergodic condition) in Stoler's sense [33]:

$$\frac{\sqrt{\bar{a}'\bar{c}' - \bar{b}'^2}}{2\pi} e^{-H(x',y')} = \frac{\sqrt{\bar{a}\,\bar{c} - \bar{b}^2}}{2\pi} e^{-H(x,y)} \tag{31}$$

Therefore, the "class" of statistical hypothesis associated to the Gaussians having the same mean, is given by the ergodic condition. This highlights the strong relationship existing among the energetic issues and the probabilistic ones] (see [71]).

2.3. Informational energy in the sense of Onicescu and uncertainty relations

For the informational energy we shall use Onicescu's relation [34, 62, 71]:

$$E = \int_{-\infty}^{\infty}\int \rho^2(x,y)dxdy \tag{32}$$

Thus, the informational energy corresponding to the normed Gaussians (20), which is subject to conditions (22), becomes

$$E\left(\bar{a},\bar{b},\bar{c}\right) = \int_{-\infty}^{\infty}\int \rho^2(x,y)dxdy \tag{33}$$

where $H(x,\ y)>0$ is a condition imposed by the existence of the integral (33).

Thus we get

$$E\left(\bar{a},\bar{b},\bar{c}\right) = \frac{\sqrt{\bar{a}\,\bar{c} - \bar{b}^2}}{2\pi} \tag{34}$$

and therefore, if H has energetic significance, it results (see [62] and [71] for details):

i. The informational energy indicates the dispersion distribution (20) because the quantity

$$A = \frac{2\pi}{\sqrt{\bar{a}\,\bar{c} - \bar{b}^2}} \tag{35}$$

is a measure of the ellipses' areas of equal probability $H(p, q)$= const., in the manner that the normed Gaussians are even more clustered the more their informational energy is higher;

ii. The class of statistical hypothesis which are specific to the Gaussians having the same mean is given by the constant value of the informational energy;

iii. The constant informational energy is equivalent to the ergodic condition;

iv. If the informational energy is constant, then the relations (21) and (34) give the egalitarian uncertainty relation

$$(\delta x)^2(\delta y)^2 = \frac{1}{4\pi^2 E^2\left(\bar{a},\bar{b},\bar{c}\right)} + cov^2(x,y) \tag{36}$$

or the non-egalitarian one

$$\delta x \delta y \geq \frac{1}{2\pi E\left(\bar{a},\bar{b},\bar{c}\right)}. \tag{37}$$

In such context we can show that the constant value of the Onicescu informational energy implies, in the case of a linear oscillator, the Planck's quantification condition.

2.4. Quantum mechanics and informational energy — Generalized uncertainty relations

[The original theory of de Broglie on the wave-corpuscle duality was developed using a theorem found in Lorentz's transformation [35]. This theorem interlinks the local horologes cyclic frequency (in each point of a spatial domain) with a progressive wave frequency in phase with the horologes. This wave gives determines the distribution of the oscillators' phases on the respective spatial domain. We desire to show that a distribution of this kind, in a true sense, can be determined without resorting to Lorentz's transformation [67].

The concept imagined by de Broglie, of equal pulsation horologes, can be evidenced by a periodic field, which is described by the local oscillators of equation

$$\ddot{Q} + \Omega^2 Q = 0 \tag{38}$$

where $Q = y + ix / m\Omega$ is the relevant complex coordinate of the field and Ω is its pulsation. The general solution of (38) can be written as [23]:

$$Q(t) = ze^{i(\Omega t+\varphi)} + \overline{z}e^{-i(\Omega t+\varphi)} \tag{39}$$

where z is a complex amplitude, $\bar{z}$ its complex conjugate and φ is a specific phase. The quantities z and $\bar{z}$ give the initial conditions, which are not the same for any point from the space. Precisely, at a time, the various oscillators corresponding to the points of the space are in different states and have different phases. A problem arises: can we apriori indicate a relationship among the parameters z, $\bar{z}$ and $e^{i(\Omega t+\varphi)}$ of the various oscillators at a given momentum? Because (39) is a solution of the equation (38) gives us an affirmative answer to this problem because (38) possesses a "hidden" symmetry that can be expressed by the homographic group: the ratio $\tau(t)$ of two solutions of the equation (38) is a solution of Schwartz's equation] [71] (see also [36, 67]).

$$\left(\frac{\tau''}{\tau'}\right)' - \frac{1}{2}\left(\frac{\tau''}{\tau'}\right)^2 = 2\Omega^2 \tag{40}$$

[This equation is invariant to the homographic transformation of $\tau(t)$: any homographic function of τ is itself a solution of (40). Since projections on the line can be characterized by the homography, we can assert that the ratio of two solutions of the equation (38) is a projective parameter for the class of the oscillators of the same pulsation from a given spatial region. Thus, one can define with ease a convenient, suitable projective parameter that should be in bi-univocal correspondence with the oscillator] [69, 71]. First, we observe a "universal" projective parameter: the ratio of the fundamental solutions of (38):

$$k = e^{2i(\Omega t+\varphi)} \tag{41}$$

Any homographic function of this ratio will be again a projective parameter [67]. Among all other, the function

$$\tau(t) = \frac{z + \overline{z}k}{1+k} \tag{42}$$

has primarily the advantage of being specific to each oscillator. But not only that: let be another function

$$\tau'(t) = \frac{z' + \overline{z}'k'}{1+k'} \tag{43}$$

which is specific to another oscillator. Since (42) and (43) are solutions of the equation (40), there exists a homographic relation between them:

$$\tau' = \frac{a'\tau + b'}{c'\tau + d'} \tag{44}$$

which, made explicit, leads to the Barbilian group equations [37]:

$$\begin{aligned} z' &= \frac{a^1 z + b^1}{c^1 z + d^1} \\ k' &= \frac{c^1 \overline{z} + d^1}{c^1 z + d^1} k \end{aligned} \tag{45}$$

[The group may be considered as a 'synchronization' group among various oscillators, a process in which the values of each take part, meaning that not only their phases, but also their amplitudes are correlated. The usual synchronization, manifested through the difference among the oscillators' phases as a whole, represents here just a particular case. Indeed, the group is involved for z, $\bar{z}$ and k, and also for (45), which indicates the fact that, indeed, the phase of k is only shifted with a value depending on the oscillator's amplitude, during passage between various members of the assembly and, moreover, the oscillator's amplitude is homographically affected.

When taking into consideration, for the group (45), the parameterization from [23], the following infinitesimal generators of the above-mentioned group will be obtained] [63]:

$$\begin{aligned} \hat{B}_1 &= \frac{\partial}{\partial z} + \frac{\partial}{\partial \overline{z}} \\ \hat{B}_2 &= z\frac{\partial}{\partial z} + \overline{z}\frac{\partial}{\partial \overline{z}} \\ \hat{B}_2 &= z^2\frac{\partial}{\partial z} + \overline{z}^2\frac{\partial}{\partial \overline{z}} + (z - \overline{z})k\frac{\partial}{\partial k} \end{aligned} \tag{46}$$

the following commutation relations

$$[\hat{B}_1, \hat{B}_2] = \hat{B}_1; \quad [\hat{B}_2, \hat{B}_3] = \hat{B}_3; \quad [\hat{B}_3, \hat{B}_1] = -2\hat{B}_2 \tag{47}$$

being involved. [Thus, a structure near-identical to group SL(2R)'s Lie algebra is shown. In consequence, the Lie algebra of the group (45) is, again, a result of group SL(2R)'s Lie algebra. Actually, as can be easily observed, the group (45) represents only another action of the group SL(2R), performed in variables z, $\bar{z}$, k] [63].

Once we fulfill the conditions of the theorem [38], the invariant functions can be found, simultaneously to the actions of the groups (5) and (46) as solutions of the equation

$$\hat{L}_i F(x,y,z,\bar{z},k)+\hat{B}_i F(x,y,z,\bar{z},k)=0,\quad i=1,2,3 \tag{48}$$

Explaining this equation by means of Equations (5) and (46) leads to their simple solution, by successive reduction, while simultaneously obtaining the invariant functions in the form

$$f(\mu,v)=const. \tag{49}$$

where μ and ν are expressed as (see [63]):

$$\begin{aligned}\mu &= \frac{-i(z-\bar{z})}{(x-zy)(x-\bar{z}y)}\\ \nu &= k\frac{x-\bar{z}y}{x-zy}\end{aligned} \tag{50}$$

ν being a unimodular complex and μ a real one. A particular class of such invariant functions is represented by linear combinations of the type

$$\bar{p}\mu = m\left(\nu+\frac{1}{\nu}\right)+2n \tag{51}$$

where m,n and $\bar{p}$ represent three arbitrary real constants.

If considering Eq (50), then Eq (51) takes the form

$$mk^{-1}z^2+2nz\bar{z}+mk\bar{z}^2\equiv\bar{p} \tag{52}$$

where the following notation has been used:

$$z=\frac{x-\bar{z}y}{\sqrt{-i(z-\bar{z})}} \tag{53}$$

We also noticed that

$$-i(z-\bar{z})>0 \tag{54}$$

Eq (52) represents a family of conical shapes from the phase space $(x,\ y)$. They represent ellipses if

$$m^2 - n^2 > 0 \tag{55}$$

a condition also fulfilled if

$$\begin{aligned} m &= \bar{Q}\sinh(2r) \\ n &= \bar{Q}\cosh(2r) \end{aligned} \tag{56}$$

where $\bar{Q}$ is a real constant and r is a real variable.

Quite an interesting case appears when z is completely imaginary, with no restriction concerning the generality value $z=i$. Thus, the quadratic form (52) may be identified with $H(p,\ q)$ from (10), resulting

$$\begin{aligned} \bar{a} &= \bar{Q}[\cosh(2r) + \sinh(2r)\cos\varphi] \\ \bar{b} &= -\bar{Q}\sinh(2r)\sin\varphi \\ \bar{c} &= \bar{Q}[\cosh(2r) - \sinh(2r)\cos\varphi] \end{aligned} \tag{57}$$

where φ is the value of k, assumed fixed. The square value of $\bar{Q}$ represents the value of the constant from (30), which determines the transitivity manifolds of the group (1) (see [63]).

The Gaussian distribution value obtained in such a manner represents only a particular case of the distribution that may occur, assuming in addition the obligation of satisfying the maximum principle of informational entropy under quadratic restrictions. The solutions of Eq (48) could be, however, much more general, being possibly selected from criteria involving group theory. In this context, the informational energy becomes

$$E(\bar{a},\bar{b},\bar{c}) = \frac{\bar{Q}}{4\pi} = const. \tag{58}$$

while the uncertainty relation (36) is

$$(\delta x)^2(\delta y)^2 = \frac{1}{\bar{Q}^2}(1 + \sinh^2(2r)\sin^2\varphi) \tag{59}$$

resulting that the concept of uncertainty is minimum only for $\varphi=0$, i.e., all the oscillators of the assembly possess the same initial phase of zero. Based on this simplified hypothesis, at any moment of time subsequent to the initial one, the uncertainty relation (59) gives up its condition of minimum, along with the assembly's covariance which differs from zero] [63].

When the creation and annihilation operators refer to a harmonic oscillator, the uncertainty relations have the form from [33, 63] with $\bar{Q}=2/\hbar$ and $\hbar$ the reduced Planck constant. In this situation, the "synchronization" is achieved through Stoler's group [33] (the parameter r is exactly equal to the frequency ratio).

Onicescu informational energy can be correlated with the standard quantum mechanics and the second quantification (which indicates its utility, for instance in NDA-NRA dynamics) [39].

2.5. Gravity and information

The structure of the group (45) is given by the equations (46) in the manner that the only non-zero structure constants are [26]:

$$C_{12}^{1} = C_{23}^{3} = -1, C_{31}^{2} = -2 \tag{60}$$

Therefore, the invariant quadratic form is given by the "quadratic" tensor of the group,

$$C_{\alpha\beta} = C_{\alpha\nu}^{\mu} C_{\beta\mu}^{\nu} \tag{61}$$

or, more explicit, by (60),

$$C_{\alpha\beta} = \begin{pmatrix} 0 & 0 & -4 \\ 0 & 2 & 0 \\ -4 & 0 & 0 \end{pmatrix} \tag{62}$$

This yields that the invariant metric of the group is given by the relation [50]

$$\frac{ds^2}{k_0^2} = \omega_0^2 - 4\omega_1\omega_2 \tag{63}$$

where k_0 is an arbitrary factor and ω_α, three differential 1-forms, which are absolutely invariant through the group.

These 1-forms have the expressions:

$$\begin{aligned} \omega_0 &= i\left(\frac{dk}{k} - \frac{dz + d\bar{z}}{z - \bar{z}}\right) \\ \omega_1 &= \bar{\omega}_2 = \frac{dz}{k(z - \bar{z})} \end{aligned} \tag{64}$$

in which case the metric (63) becomes

$$\frac{ds^2}{k_0^2} = -\left(\frac{dk}{k} - \frac{dz + d\bar{z}}{z - \bar{z}}\right)^2 + 4\frac{dzd\bar{z}}{(z - \bar{z})^2} \tag{65}$$

It should be mentioned here a property related to integral geometry: the group (45) is measurable. Indeed, it is simply transitive and, since his structure vector $C_\alpha = C^\nu_{\nu\alpha}$ is identically zero, which can be seen from (60), it means that he possesses the invariant function

$$F(z,\bar{z},k) = -\frac{1}{k(z - \bar{z})^2} \tag{66}$$

that is, the inverse of the module of the linear system's determinant obtained through the infinitesimal transformations of the group (45). Therefore, in the field variables space $(z, \bar{z}, k)$, one can build an a priori probabilities theory [40], based on the elementary probability

$$dP(z,\bar{z},k) = -\frac{dz\Lambda d\bar{z}\Lambda dk}{(z - \bar{z})^2 k} \tag{67}$$

where Λ defines the external product of the 1-forms (64).

We now analyze the metric (65): it reduces to the metric of the Lobacevski's plan in Poincaré representation [26]:

$$\frac{ds^2}{k_0^2} = 4\frac{dzd\bar{z}}{(z - \bar{z})^2} \tag{68}$$

for ω_0=0. Specifying ω_0 from (64) by the aid of the usual relations

$$\bar{z} = u + iv,\ \ k = e^{i\varphi} \tag{69}$$

it results

$$\omega_0 = -\left(d\varphi + \frac{du}{v}\right) \tag{70}$$

and so, the condition ω_0=0 becomes

$$d\varphi = -\frac{du}{v} \tag{71}$$

Since by this restriction, the metric (68) in the variables (69) reduces to the Lobacevski's one in Beltrami's representation:

$$\frac{ds^2}{k_0^2} = -\frac{du^2 + dv^2}{v^2} \tag{72}$$

the condition (71) defines a parallel transport of vectors in a Levi-Civita meaning: the application point of the vector moves on the geodesic, the vector always making a constant angle with the tangent to the geodesic in the current point. Truly, using the fact that the plan's metric is conformal Euclidean, the angle between the initial vector and the vector transported through parallelism can be calculated as the integral of the equation (see [36, 68, 71] for details):

$$d\varphi = \frac{1}{2}\left(\frac{\partial \ln E}{\partial v}du - \frac{\partial \ln E}{\partial u}dv\right) \tag{73}$$

along the transport curve. $E(u, v)$ denotes here the conformity factor of the respective metric, in our case $E(u, v)=1/v^2$. Substituting it in (73), we get (71).

Now the variables $(z, \bar{z}, k)$ can be considered as amplitudes of a gravitational field Thus, let us admit that we describe the gravitational field through the variables y_i for which we "discovered"' the metric

$$h_{ij}dy^i dy^j \tag{74}$$

in an ambient space of the metric

$$\gamma_{\alpha\beta}dx^\alpha dx^\beta \tag{75}$$

Then the field equations derive from the variational principle [41]

$$\delta \int L\gamma^{\frac{1}{2}}d^3x = 0 \tag{76}$$

relative to the Lagrange function

$$L = \gamma^{\alpha\beta} h_{ij} \frac{\partial y^i}{\partial x^\alpha} \frac{\partial y^j}{\partial x^\beta} = \gamma^{\alpha\beta} \frac{\frac{\partial z}{\partial x^\alpha} \cdot \frac{\partial \bar{z}}{\partial x^\beta}}{(z-\bar{z})^2} = \frac{" z" \bar{z}}{(z-\bar{z})^2} \tag{77}$$

In such a context, Einstein's equations with $z = i\varepsilon$ become Ernst's ones for the gravitational field of vacuum [42, 43]

$$\begin{aligned} -\frac{1}{2}(\varepsilon+\bar{\varepsilon})"\,^2\varepsilon &= "\,\varepsilon"\,\bar{\varepsilon} \\ -(\varepsilon+\bar{\varepsilon})^2 R_{\alpha\beta}(\gamma) &= \partial_\alpha \varepsilon \partial_\beta \bar{\varepsilon} + \partial_\beta \varepsilon \partial_\alpha \bar{\varepsilon} \end{aligned} \tag{78}$$

($R_{\alpha\beta}$ is here the Ricci tensor of the three-dimensional metric $\gamma_{\alpha\beta}$).

Thus, adopting as a starting point the variational principle (76), the essential goal of the analysis in the gravitational field domain is to produce metrics of Lobacevski's plan or metrics related to them. All these can be directly related to Einstein's equations (78). Moreover, by substituting the principle of independence of the simultaneous actions, in the form of linear composition in a point of various fields intensities (through the apriori invariance of fields' action with respect to a certain group), we may conceive a gravity theory that has none of the contradictions inherently and commonly present in the current theory [44]. We observe that the SL(2R) group parameters can be interpreted as field amplitudes in a supergravitation model [23] (see also [71] for details).

2.6. Extracellular vesicles convection in haptotaxis with hydrodynamical dissipation, a novel mechanism for vesicle migration

2.6.1. *On the vesicle role*

In the field of cell's biology, we call vesicles those small bags wrapped in a membrane forming part of eukaryotic cell organelles. They are involved in proteins or enzymes transport and absorption, or meet other needs of the cell. Inside the membrane bag of a vesicle, there are macromolecules which require the ability to move outside the cell walls. The membrane encircling the bag merges with the outer wall of the cell to allow such macromolecules to penetrate the wall. The vesicles are important parts of the human cells, although they are also found in other multicellular organisms [66].

Cells found in humans, plants and animals use a variety of types of vesicles, depending on the type of cell and its specific intended function. For example, one type of vesicles, lysosomes, are necessary for the process of digestion. Lysosomes contain enzymes that breakdown food cells. With food absorption, a lysosome vesicle bonds to the food holding cell and releases enzymes by a process called phagocytosis. These enzymes break down food cells into smaller parts that can be better absorbed by other cells.

Secretory vesicles are frequently associated with nerve cells in humans or animals. Their membranes sacs contain neurotransmitters. Nervous system through hormonal signals activates these components. Through the process of exocytosis, the secretory vesicle's outer membrane adheres to the nerve terminal and releases neurotransmitters in the area of the nerve endings, named the synaptic cleft. Neurotransmitters transport information from one nerve terminal to the next, across the entire central nervous system, way up to the brain [66].

Vesicles, in their role as cellular mechanism are internally appointed for transport, uptake and storage of numerous imperative bodily functions. Without these tiny bags wrapped in membranes, cells could not make the exchange of materials necessary to maintain their healthy development and other crucial processes. As a conclusion, with no vesicles, humans and other pluricellular organisms could not have existed, because the essential cellular chemical processes would have no other method to pass onto another key materials [66].

Since there is increasing support that vesicle trafficking, including the release of EVs, is a highly important process in tumorigenesis, embryogenesis and tissue remodeling, in this paragraph we present an extensive discussion on the EVs convection in haptotaxis with hydrodynamical dissipation (i.e., a novel mechanism for vesicle migration).

2.6.2. Mathematical model

Vesicles are closed membranes floating in an aqueous solution (see Fig. 1). These membranes act as a barrier that efficiently controls permeability. The vesicles mimic maybe the most primitive and mechanically flexible dividing interfaces between the inside and the outside of a cell. Generally, the fluid enclosed by the membrane is incompressible in order that the vesicle evolves at a constant volume. Moreover, the membrane exchanges no phospholipid molecules with the solution, its area remaining constant as time passes [64]. Helfrich [45] described very well the vesicle's bending energy in its equilibrium state, which is compatible with the constraints above, i.e. constant volume and area. Even if the model is relatively simple it generates various equilibrium profiles, such as, discocytes (resembling red blood cells), stomatocytes, as well as forms presenting higher topologies (such as n-genus torus) that have been also observed experimentally [46]. We identify works studying alignments of vesicle in shear flows [47], fluctuations out of equilibrium [48], lift forces [49, 50], migration of vesicle in the proximity of a substrate [51, 52] or in gravity fields [53] and also vesicle tumbling [54]. One may note several recent experiments dealing with vesicle migration [55-58].

Considering the vesicle migration, we acknowledge that hydrodynamical dissipation in the neighboring fluid as well as inside the vesicle is present, and, in principle, between the two mono-layers which may glide with respect to each other. Furthermore, during motion on the substrate the dynamics of a vesicle may be restricted not only by the hydrodynamical flow but also by bonds breaking and restoring mechanisms that occur on the substrate (see [64]). It is obvious that the slowest mechanism limits the motion. Here we focus on a situation where

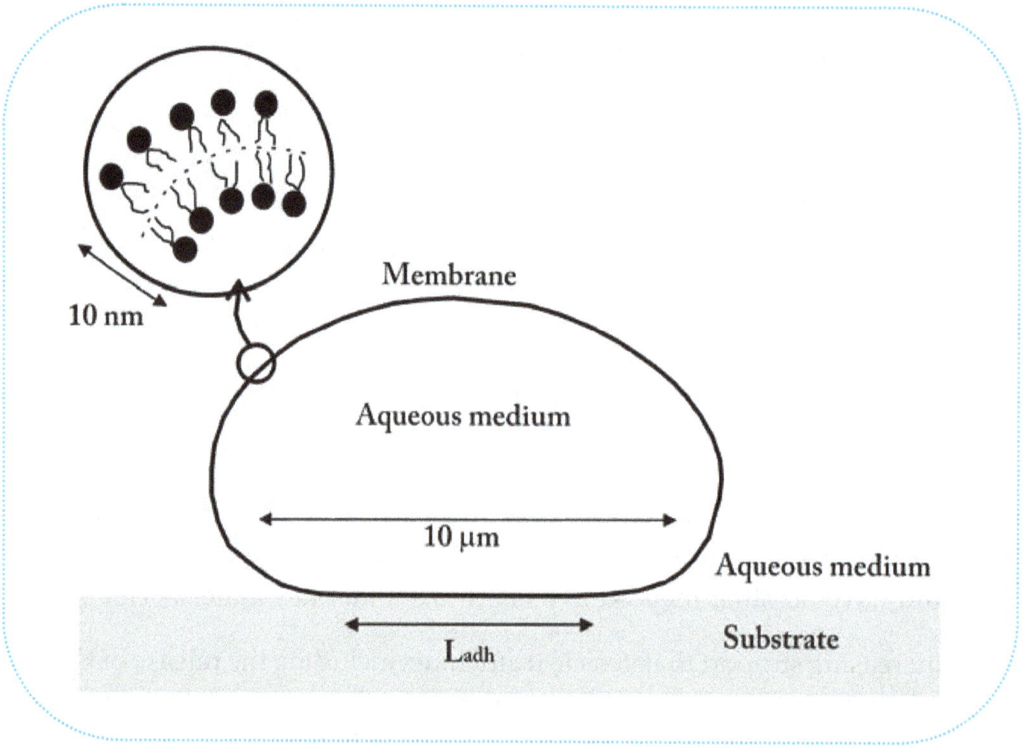

Figure 1. Schematic view of a vesicle emphasizing its microscopic structure: a bilayer of phospholipidic molecules. L_{adh} is the adherence length of the membrane

hydrodynamics are the limiting factors and we give out dissipation associated with bonds on the substrate.

Let us imagine a vesicle that initially adheres on a flat surface. We then consider an adhesion gradient along the substrate. The vesicle then moves in the direction of increasing adhesion energy (see Fig. 2) - it is named haptotaxis (a motion induced by an adhesion gradient).

A highly permeable vesicle can be pulled into a fluid without opposing any resistance (and without modifying the inner area), whereas an impermeable one would be subjected to a drag force [64]. The assumption of local impermeability is legitimate. This entails that the fluid velocity at the membrane is equal that of the membrane itself [59].

On a vesicle's scale ($R \sim 10 \mu m$) and for the expected velocities ($V \sim 1 \mu m/s$), the dissipative processes fully dominate the dynamics. The energy added instantly dissipates in various degrees of freedom. Local dissipation caused by molecular reorganization, characterized by Leslie's coefficient, is negligible with respect to the hydrodynamics modes [60].

If dissipation is dominated by bulk effects, as shown in [59], we are in the position to write down the basic governing equations for convective vesicles in a geometry depicted in Fig. 3, since we also know and it was proved the velocity field obeys Stokes equations [59].

In an original atmospheric system, the non-even distribution of ascending water droplets is determined by the interplay between solar energy-induced thermal gradients, thermal diffusivity, friction, and gravity. Ultimately, the mathematics of this model shapes the

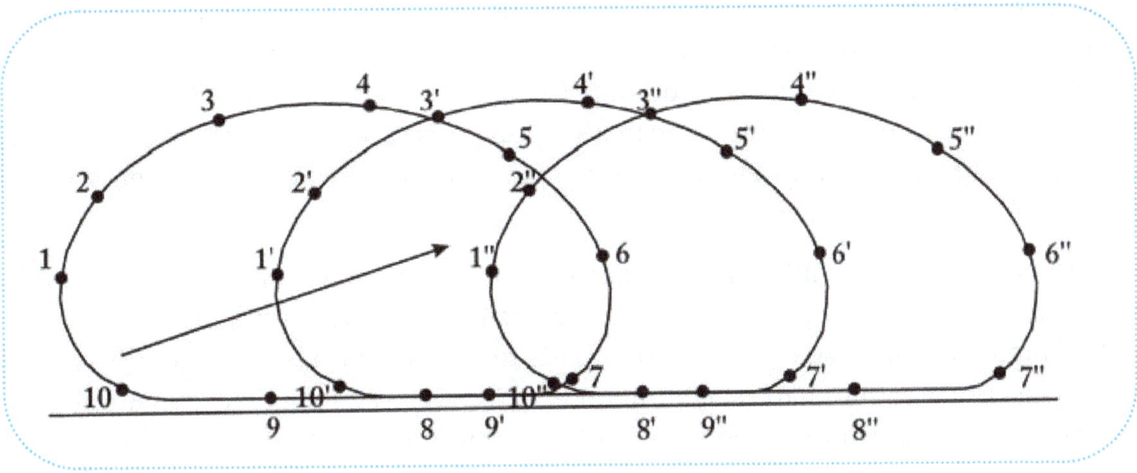

Figure 2. Stationary vesicle profiles are depicted. The vesicle is moving from the left (smaller adhesion) to the right (stronger adhesion); a few discretization points are represented and the arrow allows following one of these at three successive times. One can observe here the rolling and sliding components of the vesicle's motion.

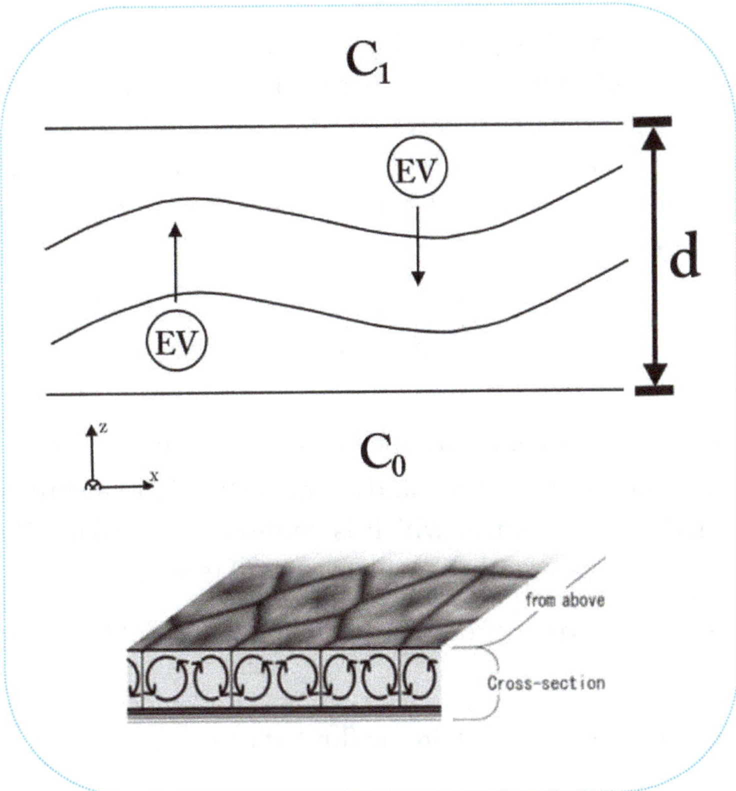

Figure 3. Convective extracellular vesicles (EVs) geometry. A fluid layer of thickness *d* of EVs, adherent on an extracellular matrix (ECM), is subjected to a gradient of concentration, where $\Delta C = C_1 - C_0 > 0$ is the difference of concentration between the front and back boundaries of the fluid layer.

umbrella-like or budding appearance of structures like cumulonimbus clouds. This model can better or uniquely describe those types of structural dynamics not explained under fractal, simple/linear and several other types of models.

Acknowledging that similar patterns occur in various biological spaces, we think that the same mathematical determinism can be ascribed. Thus, some histoarchitectural prototypic structures, like the capillary sprouting, embryologic organ, or even tumor buds of some types of cancer lesions might be in fact sculpted in that shape because gradients of molecular cues called morphogens can be deployed within the same manner water droplets can organize within nascent clouds.

Assuming that this organization also applies in biological systems, and that the EVs release can be considered among various processes organizing the budding tissue pattern, we think that the Lorenz model can govern their dynamics too. EVs would be particularly interesting as controllers of the tissue shape specification because they can include enzymatically active components (not found in conventional molecular morphogens), and thus might actively interact with the ECM fibers within their migration. Deployment of certain matrix degrading enzymes (MDEs) by EVs can modify this space while diffusing (event not produced by simple morphogens, attractive chemokines or repulsive semaphorins). This activity changes the topography of the ECM and creates spatial gradients directing the migration of subsequent EVs by haptotaxis - a mechanism better described for cell migration.

Let us consider the following thought biological experiment, equivalent to the B\'{enard experiment: a fluid layer of extracellular vesicles adherent on an ECM, in a haptotactic gradient. The fluid layer presents an unstable stratification of the potential density in a field of forces: the dense fluid is placed in front of the less dense one. We assume that in the basic state the layer of fluid of thickness d is subjected to a gradient of concentration

$$\beta = \frac{C_1 - C_0}{d} > 0 \tag{79}$$

$\Delta C = C_1 - C_0 > 0$ is the difference of concentration between the front and back boundaries of the fluid layer. The regime with the fluid at rest and a non-perturbed distribution of concentration, belongs to the thermodynamic branch, which is continuously linking the non-equilibrium stationary state ($\Delta C \neq 0$) with the equilibrium state ($\Delta C = 0$) (see Fig. 3).

We examine the evolution of a concentration fluctuation θ around the non-perturbed concentration profile $C_0(z)$.

Two dissipative processes tend to maintain the fluid at rest:

- friction (motion amortization through viscosity);
- ECM degradation subsequent to MDE's activity allowing vesicle trespassing - which lowers the concentration of the ECM, thus diminishing the forward, or advancing force.

The instability cannot be developed unless the EV is accelerated enough to overcome the effect of these dissipative processes. The gradient of concentration β which is the control parameter of this instability has to surpass a critical value β_C. Over this critical value, an organized structure of convection cells may appear.

For a one component fluid, the mass, momentum and internal energy equations are the expressions (see the fractal - nonfractal transition method [8-22]):

$$\begin{aligned}&\frac{\partial\rho}{\partial t}+\nabla\cdot(\rho\mathbf{v})=0\\&\frac{\partial(\rho\mathbf{v})}{\partial t}+\nabla\cdot(\overleftrightarrow{\Pi}+\rho\mathbf{v}\mathbf{v})=\rho\mathbf{g}\\&\frac{\partial(\rho\varepsilon)}{\partial t}+\nabla\cdot(\rho\varepsilon\mathbf{v}+\mathbf{j}_d)=-\overleftrightarrow{\Pi}\otimes(\nabla\mathbf{v})\end{aligned} \tag{80}$$

where ρ represents the mass density of the fluid, v its speed, g acceleration of a field of forces, ε the internal energy of the unit volume, and $\mathbf{j}_d$ the flux of ECM degraded by signals received from EVs. Here $\overleftrightarrow{\Pi}$ is the stress tensor and $\otimes$ denotes the product of two tensors,

$$\overleftrightarrow{A}\otimes\overleftrightarrow{B}=A_{ij}B_{ji}$$

and we use Einstein's summation convention (implicit sum over repeating indices). The stress tensor can be written

$$\overleftrightarrow{\Pi}=\overleftrightarrow{\Pi}^{e}+\overleftrightarrow{\Pi}^{v} \tag{81}$$

$\overleftrightarrow{\Pi}^{e}$ is the equilibrium part and depends on the state of the system. $\overleftrightarrow{\Pi}^{v}$ represents the non-equilibrium part and is named, viscous stress tensor. At equilibrium, this part vanishes. For an isotropic medium at rest,

$$\overleftrightarrow{\Pi}^{e}=\begin{pmatrix}p&0&0\\0&p&0\\0&0&p\end{pmatrix}=p\overleftrightarrow{I} \tag{82}$$

where p is the hydrostatic pressure. For viscous systems at non-equilibrium, the viscous stress tensor is not null. According to Eq. (81) and Eq. (82), the stress tensor will be, for homogeneous and isotropic viscous systems, at non-equilibrium

$$\overleftrightarrow{\Pi}=p\overleftrightarrow{I}+\overleftrightarrow{\Pi}^{v} \tag{83}$$

We start with the following assumptions:

a. the fluid is Newtonian; as a result the stress tensor is given by Eq. (83), where the viscous stress tensor is [8-22]

$$\Pi^{v}_{\alpha\beta} = -\eta\left(\frac{\partial v_\alpha}{\partial x_\beta} + \frac{\partial v_\beta}{\partial x_\alpha} - \frac{2}{3}\delta_{\alpha\beta}(\nabla\cdot\mathbf{v})\right) - \zeta\delta_{\alpha\beta}(\nabla\cdot\mathbf{v})$$

with coefficients η and ζ independent of velocity, the tangential (shear) and bulk viscosity, respectively;

b. ECM degrading by MDEs from EVs is described by the Fourier equation

$$\mathbf{j}_d = -\lambda\nabla C \tag{84}$$

where λ is the haptotactic coefficient;

c. haptotactic energy expansion is linear

$$\delta\rho = \rho - \rho_0 = -\rho_0\alpha\varepsilon = -\rho_0\alpha k_h(C - C_0) = -\rho_0\chi(C - C_0) \tag{85}$$

where we used the expression of the haptotactic energy

$$\varepsilon = k_h(C - C_0)$$

k_h being the haptotactic energy constant. In Eq. (85) α is the haptotactic energy expansion constant and $\chi = \alpha k_h$ is the haptotactic expansion constant;

d. the fluid satisfies a state equation: consequently, its internal energy is (up to a constant factor)

$$\varepsilon = k_b C \tag{86}$$

where k_b is the state constant;

e. in most liquids, thermal expansion is small. We choose everywhere a constant density, denoted by ρ_0, except the momentum equation.

With these approximations, the system of Eqs. (80) leads to the Boussinesq type system of equations

$$\begin{aligned}
&\nabla\cdot\mathbf{v} = 0 \\
&\rho_0\left[\frac{\partial\mathbf{v}}{\partial t} + (\mathbf{v}\cdot\nabla)\mathbf{v}\right] + \nabla p = (\rho_0 + \delta\rho)\mathbf{g} + \eta\nabla^2\mathbf{v} \\
&\frac{\partial C}{\partial t} + (\mathbf{v}\cdot\nabla)C = \frac{\lambda}{\rho_0 k_b}\nabla^2 C
\end{aligned} \tag{87}$$

where ρ is the perturbed density

$$\rho = \rho_0 + \delta\rho \tag{88}$$

The first equation (87) represents the incompressibility condition for the fluid.

Convection occurs in the fluid layer when the forward, or advancing force, resulted from energy expansion, surpasses the viscous forces. We may define now a Rayleigh type number identical with the Eqs associated to fractal - nonfractal transition [61]

$$R = \frac{|\mathbf{F}_{asc}|}{|\mathbf{F}_{visc}|} \approx \frac{\left|\frac{\delta\rho\mathbf{g}}{\rho_0}\right|}{\left|\frac{\eta\nabla^2\mathbf{v}}{\rho_0}\right|} \tag{89}$$

The density perturbation satisfies, according to Eq. (85)

$$\frac{\delta\rho}{\rho_0} \approx \chi\Delta C \tag{90}$$

On the other side, from the internal energy equation Eq. (79), it results

$$v \approx \frac{\lambda}{\rho_0 k_b}\frac{1}{d} \tag{91}$$

Replacing Eqs. (90) and (91) in Eq. (89), and taking into account Eq. (79), we get a biological Rayleigh number

$$R = \frac{\chi\beta\rho_0 k_b g}{\nu\lambda} d^4 \tag{92}$$

where $\nu = \eta / \rho_0$ is the cinematic viscosity. For the Bénard convection, the biological Rayleigh number plays the part of a control parameter. The convection occurs for

$$R > R_{critical}$$

Most of the time, R is controlled by β, the gradient of concentration.

Within a biological context, **g** can be specified by polar/linear topography of semaphorins or/ and chemokines, signals typically creating stable gradients to which EVs can respond.

We choose as reference state the rest stationary state ($v_S = 0$), for which the last two equations in the system of Eqs. (87) reduce to

$$\begin{aligned} \nabla p_S &= -\rho_S g\hat{z} = -\rho_0[1-\chi(C_S - C_0)]g\hat{z} \\ \nabla^2 C_S &= 0 \end{aligned} \tag{93}$$

where $\hat{z}$ is the versor of the vertical direction. We assume pressure and concentration varies only along the vertical direction, due to the geometry of the experiment. For concentration, the boundary conditions read

$$C(x,y,0) = C_0;\ C(x,y,d) = C_1$$

Integrating the second Eq. (93) with these boundary conditions, it results that, in the stationary reference state, the profile of the concentration in the vertical direction is linear

$$C_S = C_0 - \beta z \tag{94}$$

with β, the gradient of concentration. Replacing Eq. (94) in first Eq. (93) and integrating, we get

$$p_S(z) = p_0 - \rho_0 g\left(1 + \frac{\chi\beta z}{2}\right)z \tag{95}$$

The characteristics of the system in this state are independent of the kinetic coefficients η and λ which appear in Eqs. (87). We study the stability of the reference state using the small perturbations method. The perturbed state is characterized by

$$\begin{aligned} C &= C_S(z) + \theta(\mathbf{r},t) \\ \rho &= \rho_S(z) + \delta\rho(\mathbf{r},t) \\ p &= p_S(z) + \delta p(\mathbf{r},t) \\ \mathbf{v} &= \delta\mathbf{v}(\mathbf{r},t) = (u,v,w) \end{aligned} \tag{96}$$

As can be seen from Eqs. (96), the perturbations are functions of coordinate and time. Replacing Eqs. (96) in the evolution equations of the Boussinesq approximation Eqs. (87) and taking into account Eq. (94) and Eq. (95), we get, in the linear approximation, the following equations for the perturbations

$$\begin{aligned}
&\nabla\cdot\delta\mathbf{v}=0 && \text{a}\\
&\frac{\partial\delta\mathbf{v}}{\partial t}=-\frac{1}{\rho_0}\nabla\delta p+\nu\nabla^2\delta\mathbf{v}+g\chi\theta\hat{z} && \text{b}\\
&\frac{\partial\theta}{\partial t}=\beta w+K\nabla^2\theta && \text{c}
\end{aligned}\qquad(97)$$

where $K=\frac{\lambda}{\rho_0 k_b}$ is a coefficient. We pass to non-dimensional variables in Eqs. (97), using the transformations: $\mathbf{r}'=\frac{\mathbf{r}}{d}$; $t'=\frac{t}{d^2/K}$; $\theta'=\frac{\theta}{\left(\frac{\nu K}{g\chi d^3}\right)}$; $\delta\mathbf{v}'=\frac{\delta\mathbf{v}}{K/d}$ and $\delta p'=\frac{\delta p}{\left(\frac{\rho_0\nu K}{d^2}\right)}$.

Using the standard method from [8-22], it results the biological Lorentz system.

2.6.3. Validity of theoretical model

Some results are evident:

- we build the first Lorenz model for extracellular vesicles migration;
- in [50], and similar other references ([64] etc.), the EVs behavior, under shear flow close to a substrate, was proved to be quite similar to the one encountered in two dimensional simulations, so we are confident that the 2D assumptions captures the essential features of the 3D EVs;
- different control parameter values for the Lorenz system can create shape distributions similar to the cordonal appearance of fingerprints (see Fig. 4, A), or complex skin tissue tiles like scale appendages in the amphibian covering (see Fig. 4, B);
- the biological thought experiment equivalent to Bénard's experiment, involving a fluid layer of extracellular vesicles adherent to an extracellular matrix, in a haptotactic gradient can be *checked experimentally* today to a high degree of accuracy. We think that suitable test systems would be the *embryological ones* (i.e., the development of branched vessels in membranes - avian eggshell membranes, serous membranes of the peritoneal cavity; or the budding development of lung alveoli, or of fingerprints), and, similar, *inflammatory ones* (i.e., the emergence of neoangionetic vessels driven by inflammatory proximities) - all of which apparently start as point like spots displayed in a comb-like appearance along a rectilinear or arched origin;
- we analyze the problem of EVs migration in haptotaxis, though most of the reasoning applies to chemotaxis (migration of cells biased towards a gradient of diffusible MDEs) as well as to a variety of driving forces - all of which include the possibility to specify an active parameter value within the model;
- the resulted system of equations exhibits complex behavior, hard to control, the two occurring convective rolls: either going in one direction, or in the opposite one - means patterning the EVs spreading.

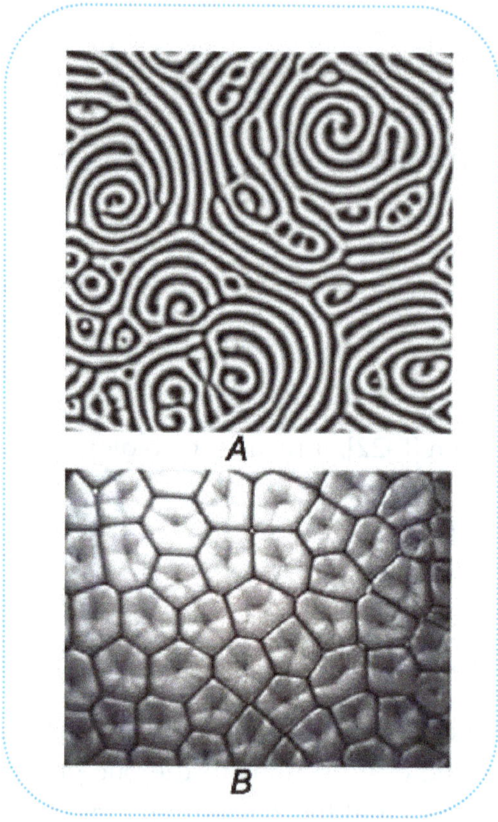

Figure 4. Bénard-Rayleigh model patterns representative for biological instances: A) for fingerprint like distribution of skin cells, B) for fish or amphibian scales.

3. Conclusions

Considering the above, we can write the following conclusions :

i. We establish a relationship between the SL(2R) group and the canonic formalism. It particularly results that all invariant functions on the SL(2R) group will be functions on the hamiltonian and on the Gausssian;

ii. We establish the statistical significations for coefficients of the hamiltonian by using Shannon's maximum informational entropy principle. Any statistical assumption is specified by particularly selecting the hamiltonian coefficients and the class of all these hypothesis by the transitivity manifolds of the group. In this manner, if the hamiltonian has energy significance, then through the transitivity manifolds, the motions of a representative point from the phase space on a surface of constant energy are in the same time on a surface of constant probability density (the ergodic hypothesis). Therefore, the class of the statistic hypothesis (which are characteristic to the Gaussians of the same average) is given by the ergodic hypothesis. In this way, we establish a fundamental relationship between energy and probability;

iii. We prove that informational energy (in the sense of Onicescu) is a measure of the dispersion of a distribution. The class of the statistic hypothesis that are characteristic to the Gaussians of the same average is characterized by the constant value of the informational energy. In addition, it is equivalent to the ergodic hypothesis. For a constant value of the informational energy, we obtain uncertainty egalitarian relationships and, particularly, for the linear harmonic oscillator, we show that the informational energy is quantified;

iv. Assuming that de Broglie's theory is materialized through a periodic field in a complex coordinate, we prove that it has a "hidden symmetry", which is expressed by the homographical transformations group in three parameters. This group (also an achievement of SL(2R)) functions as a synchronization group both in phase and in amplitude, among the oscillators of the same ensemble. The simultaneous invariance related to two different achievements of SL(2R) implies (integrally through the invariant functions on the groups) an uncertainty relation in egalitarian form and the Stoler group of synchronization among oscillators from different ensembles (i.e., the second quantification when the creation and annihilation operators refer to a harmonic oscillator);

v. The synchronization group among the oscillators of the same ensemble admits three differentiable 1-forms and one differentiable 2-form which is absolutely invariant on the group. The existence of a parallel transport in Levi-Civita's sense, in which case the 2-form in Lobacevski's metric form in Poincare representation, implies through a variation principle, equations of Ernst type for the gravitational field of vacuum;

vi. Complex measures in the study of certain physical systems dynamics need the use of a space-time endowed with a special topology, namely, the fractal space-time [6, 7] (see also [71] for details).

Author details

Maricel Agop[1,2*], Alina Gavriluţ[3], Călin Buzea[4], Lăcrămioara Ochiuz[5], Dan Tesloianu[6], Gabriel Crumpei[7] and Cristina Popa[8]

*Address all correspondence to: m.agop@yahoo.com

1 Lasers, Atoms and Molecules Physics Laboratory, University of Science and Technology, Lille, France

2 Department of Physics, "Gh. Asachi"' Technical University, Iaşi, România

3 Department of Mathematics, "Al.I. Cuza" University from Iaşi, Romania

4 National Institute of Research and Development for Technical Physics, Iaşi, România

5 Faculty of Pharmacy, Department of Pharmaceutical Technology University of Medicine and Pharmacy "Gr.T. Popa", Iași, România

6 IV Internal Medicine Department, Clinical Emergency Hospital "Sf. Spiridon", University of Medicine and Pharmacy "Gr.T. Popa", Iași, România

7 Psychiatry, Psychotherapy and Counselling Center, Iași, România

8 Dentistry Department, University of Medicine and Pharmacy "Gr.T. Popa", Iași, România

References

[1] Flake G. W., *The Computational Beauty of Nature*, MIT Press, Cambridge, MA, 1998.

[2] Mitchell M., *Complexity: A Guided Tour*, Oxford University Press, Oxford, 2009.

[3] Winfree A. T., *The Geometry of Biological Time*, Springer 2nd edition, New York, 2000.

[4] Bennett C. H., *How to define complexity in physics, and why*, in W. H. Zurek (editor), Buldyrev S. V., Parshani R., Paul G., Stanley H. E., Haylin S., *Catastrophic cascade of failures in independent networks*, Nature 464, 2010

[5] Badii R., Politi A., *Complexity: Hierarchical Structure and Scaling in Physics*, Cambridge University Press, Cambridge, 1997.

[6] Nottale L., *Fractal space-time and microphysics: towards a theory of scale relativity*, World Scientific, Singapore, 1993.

[7] Nottale L., *Scale relativity and fractal space-time – a new approach to unifying relativity and quantum mechanics*, Imperial College Press, London, 2011.

[8] Agop, M., Dimitriu, D., Vrăjitoriu, L., Boicu, M., *Order to chaos transition in plasma via non-differentiability: Experimental and Theoretical Investigations*, Journal of the Physical Society of Japan 83, 054501, 2014.

[9] Agop M., Forna N., Casian-Botez I., Bejinariu C., *New theoretical approach of the physical processes in nanostructures*. Journal of Computational and Theoretical Nanoscience, 5 (4), 483-489, 2008.

[10] Agop M., Murguleț C., *El Naschie's epsilon (infinity) space-time and scale relativity theory in the topological dimension D=4*, Chaos Solitons & Fractals. 32 (3), 1231-1240, 2007.

[11] Agop M., Nica P., Gîrțu M., *On the vacuum status in Weyl-Dirac theory*, General Relativity and Gravitation, 40 (1), 35-55, 2008.

[12] Agop M., Păun V., Harabagiu A., *El Naschie's epsilon (infinity) theory and effects of nanoparticle clustering on the heat transport in nanofluids*, Chaos, Solitons and Fractals, 37 (5), 1269-1278, 2008.

[13] Casian-Botez I., Agop M., Nica P., Păun V., Munceleanu G.V., *Conductive and convective types behaviors at nano-time scales*, Journal of Computational and Theoretical Nanoscience, 7, 2271-2280, 2010.

[14] Ciubotariu C., Agop M., *Absence of a gravitational analog to the Meissner effect*. General Relativity and Gravitation, 28 (4), 405-412, 1996.

[15] Colotin M., Pompilian G. O., Nica, P., Gurlui S., Păun V., Agop M., *Fractal transport phenomena through the scale relativity model*, Acta Physica Polonica A, 116 (2), 157-164, 2009.

[16] Gottlieb I., Agop M., Jarcău M., *El Naschie's Cantorian space-time and general relativity by means of Barbilian's group. A Cantorian fractal axiomatic model of space-time*, Chaos, Solitons and Fractals, 19 (4), 705-730, 2004.

[17] Gurlui S., Agop M., Nica P., Ziskind M., Focșa C., *Experimental and theoretical investigations of transitory phenomena in high-fluence laser ablation plasma*, Phys. Rev. E 78, 026405, 2008.

[18] Gurlui S., Agop M., Strat M., Băcăiță S., *Some experimental and theoretical results on the anodic patterns in plasma discharge*, Physics of Plasmas, 13 (6), 063503, 2006.

[19] Nedeff V., Bejenariu C., Lazăr G., Agop M., *Generalized lift force for complex fluid*, Powder Technology, 235, 685-695, 2013.

[20] Nedeff V., Moșneguțu E., Panainte M., Ristea M., Lazăr G., Scurtu D., Ciobanu B., Timofte A., Toma S., Agop M., *Dynamics in the boundary layer of a flat particle*, Powder Technology, 221, 312-317, 2012.

[21] Nica P., Agop M., Gurlui S., Bejinariu C., Focșa C., *Characterization of aluminum laser produced plasma by target current measurements*, Japanese Journal of Applied Physics, 51, DOI: 10.1143/JJAP.51.106102, 2012.

[22] Nica P., Vizureanu P., Agop M., Gurlui S., Focșa C., Forna N., Ioannou P. D., Borsos Z., *Experimental and theoretical aspects of aluminum expanding laser plasma*, Japanese Journal of Applied Physics, 48 (6), DOI: 10.1143/JJAP.48.066001, 2009.

[23] Berners-Lee T. and Kagal L., *The fractal nature of the semantic web*, AI Magazine, 29(8), 2008, 29--34.

[24] Gardiner J., et al., *The fractal nature of the brain*, NeuroQuantology, 8(2), 2010, 137-141.

[25] Agop M., Mazilu N., *Skyrmions: A Great Finishing Touch to Classical Newtonian Philosophy*, Nova Publishers, New York, 2012.

[26] Casanova G., *L'Algebre Vectorielle*, Mir, Moscou, 1979.

[27] Arnold V.I., *Metode Matematice ale Mecanicii Clasice*, Ed. Tehnică, București, 1980.

[28] Hamermesh M., *Group Theory and its Applications to Physical Problem*, Dover Publication, Inc. New York, 1989.

[29] Alipour M., Mohajeri A., *Onicescu information energy in terms of Shannon entropy and Fisher information densities*, Molecular Physics: An International Journal at the Interface Between Chemistry and Physics, Vol. 110, Issue 7, 403-405, 2012.

[30] Berners-Lee T., Kagal L., *The Fractal Nature of the Semantic Web*, AI Magazine, 29 (8), 29-34, 2008.

[31] Jaeger G., *Fractal states in quantum information processing*, Physics Letters A, Vol. 358, Issues 56, 373-376, 2006.

[32] Agop M., Buzea C., Buzea C.Gh., Chirilă L., Oancea S., *On the information and uncertainty relation of canonical quantum systems with SL(2R) invariance*, Chaos, Solitons & Fractals, Vol. 7, Issue 5, 659-668, 1996.

[33] Agop M., Griga V., Ciobanu B., Buzea C., Stan C., Tatomir D., *The uncertainty relation for an assembly of Planck-type oscillators. A possible GR-quantum mechanics connection*, Chaos, Solitons & Fractals, Vol. 8, Issue 5, 809-821, 1997.

[34] Agop M., Melnig V., *L'énergie informationelle et les relations d'incertitude pour les systèmes canoniques SL(2R) invariants*, Entropie, no. 188/189, 119-123, 1995.

[35] Stoler D., *Equivalence Classes of Minimum Uncertainty Packets*, Phys. Rev. D, 3217-3219, 1970.

[36] Onicescu O., *Energie informationnelle*, C. R. Acad. Sci. Paris A 263 (1966), 841-842.

[37] Broglie L. de, *La Thermodynamique de la particule isolée*, Gauthier-Villars, Paris, 1964.

[38] Mihăileanu N., *Complements of Geometry: Analytical, Projective and Differential*, Ed. Didactică și Pedagogică, București, 1972.

[39] Barbilian D., *Riemannsche Raum Cubischer Binärformen*, Comptes Rendus de l'Academie Roumaine des Sciences, Vol. 12, p. 345.

[40] Stoka M.I., *Géométrie Intégrale*, Memorial des Sciences Mathématiques, Gauthier-Villars, Paris.

[41] Flores-Gallegos Nelson, *Shannon informational entropies and chemical reactivity*, in Advances in Quantum Mechanics, Edited by Paul Bracken, Intech 2013, Rijcka, Croatia, 683-706.

[42] Jeans J.H., *The Dynamical Theory of Chaos*, Dover Publication, New York, 1954.

[43] Misner C.W., *Harmonic Maps as Models for Physical Theories*, Physical Review D 18, 4510-4524.

[44] Ernst F.J., *New Formulations of the Axially Symmetric Gravitational Field Problem I*, Phys. Rev., 167, 1175-1178, *New Formulation of the Axially Symmetric Grvavitational Field Problem II*, Phys. Rev., 168, 1415-1417.

[45] Ernst F.J., *Exterior Algebraic Derivation of Einstein Field Equations Employing a Generalized Basis*, Journal of Mathematical Physics, 12, 2395-2397, 1971.

[46] Misner C.W., Thorne K.S., Wheeler J.A., *Gravitation*, W.H. Freeman and Company, San Francisco, 1973.

[47] Helfrich W., *Elastic properties of lipid bilayers-theory and possible experiments*, Z. Naturforsch. C 28, 693-703, 1973.

[48] Lipowsky, R. and Sackmann, E., *Structure and Dynamics of Membranes*, Handbook of Biological Physics. R. Lipowsky and E. Sackmann, editors. Elsevier, North-Holland, 1995.

[49] Kraus, M., Wintz, W., Seifert, U. and Lipowsky, R., *Fluid vesicles in shear flow*, Phys. Rev. Lett. 77, 3685-3688, 1996.

[50] Prost, J. and Bruinsma, R., *Shape fluctuations of active membranes*, Europhys. Lett. 33(4), 321-326, 1996.

[51] Cantat, I. and Misbah, C., *Dynamics and similarity laws for adhering vesicles in haptotaxis*, Phys. Rev. Lett. 83, 235-238, 1999.

[52] Sukumaran, S. and Seifert, U., *Influence of shear flow on vesicles near a wall: A numerical study*, Phys. Rev. E 64, 011916, 2001.

[53] Durand, I., Jönson, P., Misbah, C., Valance, A. and Kassner, K., *Adhesion-induced vesicle propulsion*, Phys. Rev. E 56, 3776, 1997.

[54] Cantat, I. and Misbah, C., *Lift Force and Dynamical Unbinding of Adhering Vesicles under Shear Flow*, Phys. Rev. Lett. 83, 880-883, 1999.

[55] Kern, N. and Fourcade, B., *Vesicles in linearly forced motion*, Europhys. Lett. 46(2), 262-267, 1999.

[56] Biben, T. and Misbah, C., *An advected-field method for deformable entities under flow*, Euro. Phys. J. B 29, 311-316, 2002.

[57] Nardi, J., Bruinsma, R. and Sackmann, E., *Vesicles as osmotic motors*, Phys. Rev. Lett. 82, 5168-5171, 1999.

[58] Abkarian, M., Lartigue, C. and Viallat, A., *Motion of phospholipidic vesicles along an inclined plane. Sliding and rolling*, Phys. Rev. E 63, 041906, 2001.

[59] Lortz, B., Simon, R., Nardi J. and Sackmann, E., *Weakly adhering vesicles in shear flow*, Tanktreading and anomalous lift force, Europhys. Lett. 51, 468, 2000.

[60] Abkarian, M., Lartigue C. and Viallat, A., *Tank treading and unbonding of deformable vesicles in shear flow. Determination of the lift force*, Phys. Rev. Lett. 88, 068103, 2002.

[61] Cantat, I., Kassner K. and Misbah, C., *Vesicles in haptotaxis with hydrodynamical dissipation*, Eur. Phys. J. E 10, 175-189, 2003.

[62] Agop, M., Gavriluţ, A., Crumpei, G., Doroftei, B., Informational Non-Differentiable Entropy and Uncertainty Relations in Complex Systems, Entropy, 2014, 16(11), 6042-6058.

[63] Agop, M., The uncertainty relation for an assembly of Planck-type oscillators. A possible GR-quantum mechanics connection, Chaos, Solitons and Fractals, 1997, Vol. 8, No. 5, 809-821.

[64] ***http://perso.univ-rennes1.fr

[65] Wilhelm, H., Hydrodynamic Model of Quantum Mechanics, Physical Review D, vol. 1, no. 8, 2278-2285.

[66] *** http://www.wisegeek.com

[67] Gottlieb, I., *El Naschie's Cantorian frames, gravitation and quantum mechanics*, Chaos, Solitons and Fractals, 2005, vol. 24, Issue 2, 391-405.

[68] Agop, M., Ciobanu, G., Zaharia, L., *Cantorian* $\varepsilon^{(\infty)}$ *space-time, frames and unitary theories*, Chaos, Solitons and Fractals, 2003, vol. 15, no. 3, 445-453.

[69] Agop, M., Abacioaie, D., El Naschie's $\varepsilon(\infty)$ space-time, interface between Weyl-Dirac bubbles and Cantorian fractal superstring, Chaos, Solitons and Fractals, 2007, vol. 34, Issue 2, 235-243.

[70] ***http://pirsquared.org

[71] Agop, M., Gavriluţ, A., Rezuş, E., Implications of Onicescu's informational energy in some fundamental physical models, International Journal of Modern Physiscs B, DOI: 10.1142/S0217979215500459, 2015.

[72] *** http://www.canaryjournal.org

3

Implications of the "Subquantum Level" in Carcinogenesis and Tumor Progression via Scale Relativity Theory

Daniel Timofte, Lucian Eva, Decebal Vasincu,
Călin Gh. Buzea, Maricel Agop and Radu Florin Popa

1. Introduction

The last 25 years witnessed tremendous achievements in cancer diagnose and treatment. Technology currently permits small size tumors (like breast cancers) diagnosis and treatment, ductal cancer in situ currently including 25...30% of all freshly diagnosed breast cancers at the majority of medical centers [1]. Thus, early detection now allows the understanding of growth patterns. Surgeons are in the front line of technological and basic scientific medical advances. Current ideas, such as the physiological characteristics of shock, organ transplantation, antisepsis, wound healing, or sequence medical care, are cast by surgical investigators.

The field of mathematics suffered an identical evolution. Revolutionary mathematical branches such as topology, fractals, chaos theory, and development of nonlinear descriptive strategies have provided mathematicians new inventive tools to create growth models and to behaviors at the small environmental level [2, 3]. Growth, angiogenesis [4], cell-to-cell adhesion [5], hydrogen ion concentration regulation and drug delivery [6] can now systematically be described using specific formulas. From a clinical viewpoint several of these formulas could seem simple, however they put together a very important foundation for descriptive insight.

What is currently lacking is a connection between these two naturally and mutual analysis endeavors. For oncology surgeons, the ability to mathematically analyze and predict patterns of growth provides precise techniques that are beneficent for both current and future therapies. For mathematicians, defining the clinical factors essential for growth development and metastasis can provide realistic insight into these biological processes, successively allowing the event of correct, clinically relevant mathematical formulas. In almost every dedicated medical institutions, the teams that have comprehensive cancer are being led by surgeons. Mathematics that can be applied in the oncology field provide a chance to expand the leadership role of the surgeons and to raise awareness about the significance of understanding growth behavior and, also improve cancer treatments.

The present study aims at defining a new concept of carcinogenesis and tumor progression. Consequently, we use the natural 'environment' where malignant tumors grow, space(-time) with non-integer fractal dimension, questing for further applications of the newly discovered and intriguing phenomenon of tumor self-seeding by circulating cancer cells (CTC). More precisely, we assume that the metastatic tumor cells move (through the systemic circulation, yet not necessarily only there) as a coherent wave, or even more precisely, a chemically pumped travelling laser wave with oxygen. The extracellular matrix (ECM) and in particular, the tumor microenvironment (TME) are assumed as non-differential media endowed with holographic properties and may be good candidates for "recording" materials. As a result, the tumor self-seeding by CTC may be proved mathematically, the fact that the CTC returning to the initial tumor site and fueling the primary tumor growth or even grow a new tumor is a particular case of complete holography (i.e. a hologram which does not represent only the virtual object's image, but it becomes the very object - which we believe, is a characteristic of the living organisms). We believe our findings may provide new opportunities to set up new targeted therapies that may slow down or even prevent tumor progression.

According to the Pribram-Bohm's holographic theory (http://en.wikipedia.org/wiki/Holonomic_brain_theory) of the brain, the intercellular and intracellular communication implies the existence of a fractal medium equivalent to the vacuum between the elementary particles. Since vacuum dynamics is studied using quantum mechanics, it is only natural that the status of the fractal medium implies the use of a quantum type mathematical formalism (i.e. Scale Relativity theory either in its Schrödinger type representation or the fractal hydrodynamic one) applicable to different resolution scales (mesoscopic for intercellular communications or nanoscopic for intracellular ones). In Nottale's interpretation, each resolution scale is characterized by a Planck type constant $\bar{h}$

$$\bar{h} = \hbar \left(\frac{dt}{\tau} \right)^{(2/D_F)-1}$$

where $\hbar$ is the standard Planck's constant, dt is time's resolution scale, τ the time's reference scale and D_F is the fractal dimension of the motion curve.

2. The biology of cancer

2.1. Cancer, what should be noticed

Cancer or malignant neoplasm is a class of diseases that rises from the anomalous behavior of normal tissue. Cancer cells are aberrant cells which have acquired malignant traits such as uncontrolled growth (cells continuously proliferate), tissue invasion (they intrude into normal tissue and destroy it) and metastasis (they spread outside the location of the body where they were originally generated). Additionally, the term tumor or neoplasm is used to indicate an abnormal swelling of tissue caused by an excessive cell proliferation.

A tumor can be of benign or malignant nature, while benign tumors are self-limiting, do not express patterns of invasion, and they do not metastasize, malignant tumors do possess all these characteristics. The term malignant tumor is also used as synonym for cancer, although some cancers, such as leukemia, do not form tumors.

Cancer cells develop these malignant features because of genetic mutations, accumulated during the organism lifetime. Cancer is in fact a multi-step chance process that transforms a normal cell into a tumor cell, after having collected a set of 5...8 crucial genetic alterations [7-9] as schematically shown in Fig. 1.

A newborn malignant cell, expressing aberrant traits, can lead to the formation of cancer and, in most of the cases, of a tumor. Without treatment, the destructive behavior of such colony of cells is usually lethal for the patient. The probabilistic nature of this disease and the increase in life expectancy had made cancer the second cause of death in the industrialized countries (see any cancer statistics). Nevertheless, cancer it is not a modern disease and it was known since the antiquity: Egyptians of the New Kingdom [10], Greeks [11] and Romans [12] accurately described medical treatments for tumor removal. It is only within the last two centuries however, that due to the higher standards of living, cancer has become one of the main life-threatening diseases.

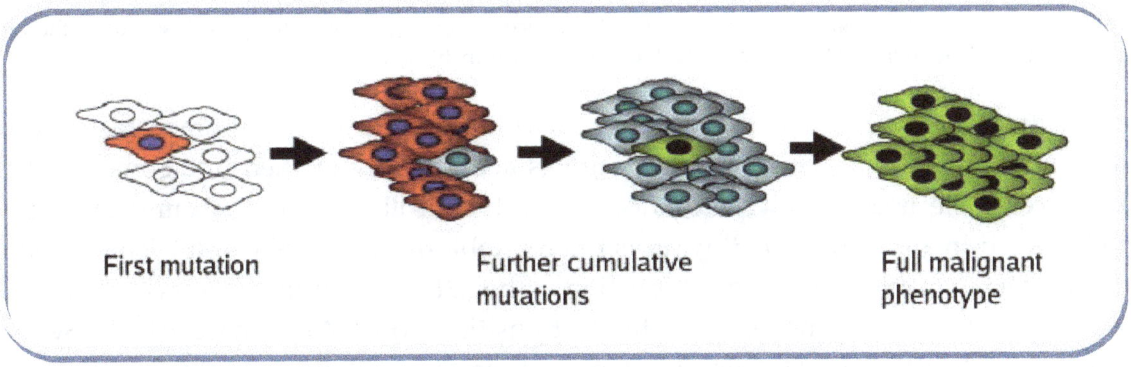

Figure 1. Acquisition of the tumorigenic phenotype by a population of normal cells through multiple genetic mutations.

2.2. Distinguishing traits of cancer

The tumorigenic properties, generically discussed in the previous section, have shown to be common to almost all cancers. They have been studied since the dawn of cancer research and they can be enumerated and defined with a relatively high accuracy. These hallmarks are a set of characteristic traits typical of cancer cells that are essential for the formation of a macroscopic malignant neoplasm [13]:

Self-Sufficiency in Growth Signals. All cells communicate through signals. A biological signal is, in most of the cases, a protein able to deliver a particular piece of information by binding uniquely to specific receptors on the cell surface. Normal cells need mitogenic growth signals to proliferate (signals that allow and stimulate cell proliferation). Those signals are regulated by the homeostasis of the tissue and they guarantee a correct balance between cell proliferation and death, according to the needs of the organism. In order to lead to cancer, tumor cells may develop the ability of self-generating such signals in one way or another. One possible way is a genetic aberration in one of the fundamental genes responsible for the building of the signaling pathway, for instance the RAS oncogene [9,14]. As consequence, the associated component of the signaling system would become constitutively active and hence, independent by the signal molecule. A second option is the self-production of growth factors that would stimulate growth by paracrine signaling, where a cells stimulates the neighbors and vice-versa or even autocrine signaling, when the cell stimulates its own receptors as shown in Fig. 2.

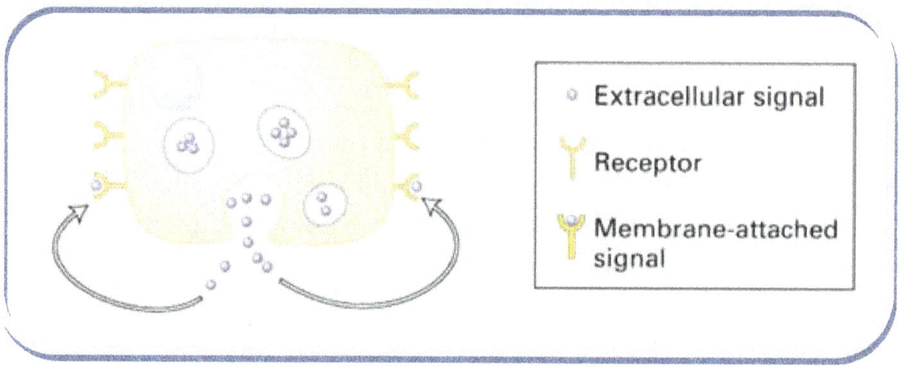

Figure 2. Example of self-signaling (autocrine): the cell produces its own growth factors which stimulate the growth receptors on the surface (Dr. W.H. Moolenaar, Netherlands Cancer Institute).

Insensitivity to Antigrowth Signals. As counterparts of growth factors, homeostasis employs growth inhibiting signals as well. These signals act similarly to their antagonists but they promote cell cycle arrest or cell quiescence, rather than proliferation. An example of a crucial gene involved in anti-growth pathways is the retinoblastoma protein (pRb). The retinoblastoma protein is capable of altering the function of the E2F transcription factors and control the expression of the bank of genes essential for the transition from GAP-1 phase to DNA Synthesis phase of the cell cycle [15]. The disruption of such pathway results in the insensitivity of the cell to anti-growth signals.

Evading Apoptosis. Apoptosis is a mechanism of controlled cell death. Through special signals, a cell has the capacity of terminating itself in a highly regulated way. A normal cell dying by apoptosis undergoes a sequence of events such as condensation, fragmentation and phagocytosis. This avoids the cell to free potentially dangerous enzymes and proteins stored inside its cytoplasm and its nucleus. During apoptosis the cell membrane is kept intact while, in 30...120 minutes the cell is fragmented in small parts or apoptotic bodies, still protected by pieces of membrane. Those cell leftovers are successively phagocytated by macrophages within the next 24 hours [16]. Apoptosis is a common mechanism of cell death and takes part in the homeostasis of a healthy organism as well as in its embryogenesis and in its morphogenesis. When any cell violates such homeostasis, an apoptotic signal is delivered to it. Therefore, in order for cancer cells to develop into a malignant lesion, it is necessary to deactivate apoptotic signal pathways. A mutation in the p53 tumor suppressor gene (TSG) is one of the most common ways to acquire resistance to apoptosis because p53 regulates the whole signaling process of programmed cell death. Indeed, more than 50% of human cancers carry a mutation in the p53 tumor suppressor gene [17].

Limitless Replicative Potential. Even with all the anti-growth and anti-apoptosis pathways triggered off, a cell could not generate a vast population able to form a tumor. That is because of the intrinsic proliferation limit of all mammalian cells. All chromosomes have an ending cap called telomere, a T-loop non-coding DNA sequence (2...50 Kb) that prevents the end of the chromosomes from attaching to other genetic material. At every mitosis the cell loses a small part of its telomeres because of the impossibility for DNA duplication enzymes, for instance DNA-polymerase, to continue working until the very end of the genome (Fig. 3). This limitation is due to the fact that enzymes like DNA-polymerase always move in the 5′...3′ direction of the DNA sequence, so when the side of the replication is opposite, a small part of the genome is lost. The shortening of the telomeres induces cell senescence, a state of cellular elderly where division no longer occurs. This avoids genetically unstable cells to replicate. Senescence starts after the so called Hayflick limit [18] of about 50 cell divisions. In cancer cells instead, the disabling of the pRb and the p53 pathways allows unlimited replication, until the point when the telomeres are completely absent. Once having entirely consumed the telomeres, the cell population is believed to undergo a phase of massive genomic instability, causing extended cell death. The high selective pressure induced by this crisis may permit specific resistant clones to emerge (Fig. 4). Those survivor cells would be immortalized (unlimited proliferative potential) by finding ways to maintain their telomeres long enough. A possible way is the over expression of the telomerase gene [19] which appears to take place in 85...90 % of cancers. Telomerase is a telomere-rebuilding enzyme normally expressed in germ line cells and stem cells, in which immortalization is an essential feature. Once immortalized, malignant cells have made a further step towards the formation of cancer.

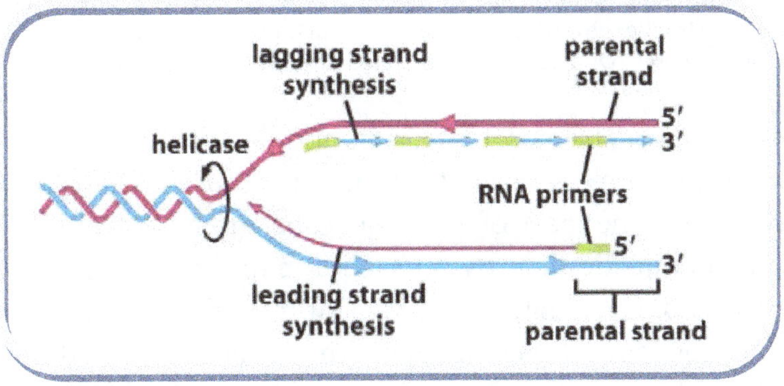

Figure 3. Illustration of the end replication problem: at both sides of the copying, the leading DNA strand has lost part of the telomeric sequence, which stops at the 5′ end of the parental strand, whereas the lagging strand results completed until the very end (Dr. R. Beijersbergen, Netherlands Cancer Institute).

Sustained Angiogenesis. In order for a cell to survive normally, it must rely within 100 μm from a capillary blood vessel [13]. For this reason, the initial exponential growth of a newborn malignant neoplasm causes a shortage of nutrients among cancer cells. Local pre-existent vascularisation is never enough to sustain growth for more than 10^8 cells. The colony must therefore develop angiogenesis-triggering capabilities [20, 21]. Angiogenesis is the process of formation of new blood vessels in response to a stimulus secreted by poor vascularised tissues. Angiogenesis is important for the organism morphogenesis and even later maintains the correct supply of nutrients for all tissues. Fast growing cells, such as cancer cells, start soon to starve, and have the need of additional blood supply in order to keep expanding. A possible solution is the production by cancer cells of vascular endothelial growth factors (VEGF) and fibroblast growth factors (FGF1/2) which bind to the transmembrane receptors of endothelial cells (cells covering the interior surface of blood vessels) stimulating their growth towards the signal concentration gradient [22]. Angiogenesis is the principal mechanism that transforms a microscopic malignancy into a macroscopic tumor and, also in the later stages, it is necessary for a lesion to grow and sustain itself. This implies that angiogenesis is an important target for anti-cancer drugs like thrombospondin-1 [23] and bevacizumab [24], also known as avastin.

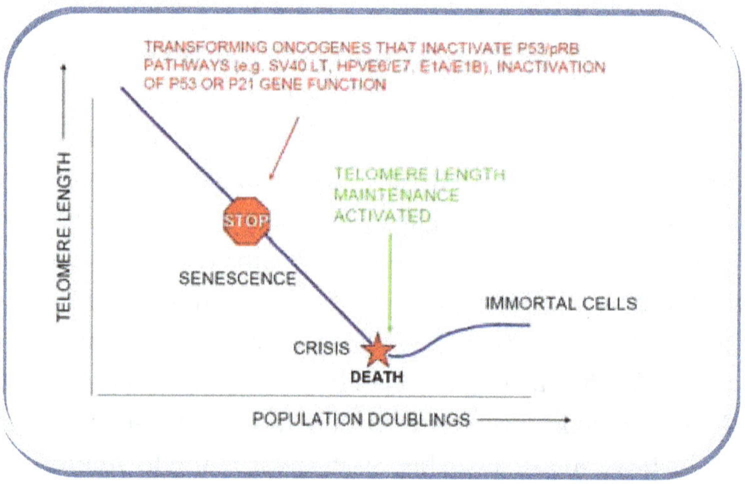

Figure 4. The progressive shortening of the telomeres leads to a massive cell death due to the induced genomic instability (death by genomic catastrophe while duplicating). From such process of intense genetic mutation and selection an immortalized clone could emerge (Dr. R. Beijersbergen, Netherlands Cancer Institute).

Tissue Invasion and Metastasis. The most dangerous and destructive features of cancer are tissue invasion and the consequent metastasis. Its ability of forming distant colonies or metastases all over the body represents the cause of 90% of all cancer related deaths [25]. Normal cells are usually unable to travel outside their own tissue due to their necessity to be anchored and reside among similar cells. An eventual detachment from the extracellular matrix or ECM (a complex structure of proteins and specific cells forming the tissue scaffold and microenvironment – see Sec. 6.1) would occur in a form of apoptosis called anoikis [26]. Contrary to their normal counterparts, cancer cells are able to survive the loss of anchorage, to travel through the vascular system and form distant tumors elsewhere (Fig. 5). The traits expressed by invasive and metastatic cancer cells are principally loss of cell-to-cell adhesion, anchorage-independence, chemotaxis (migration towards a diffusible substance gradient), haptotaxis (migration towards a non-diffusible substance gradient) and production of matrix degrading enzymes (*e.g.* Matrix metalloproteinase) which cleave the extracellular matrix [27-29] making space for invasion and freeing growth and angiogenic factors trapped inside.

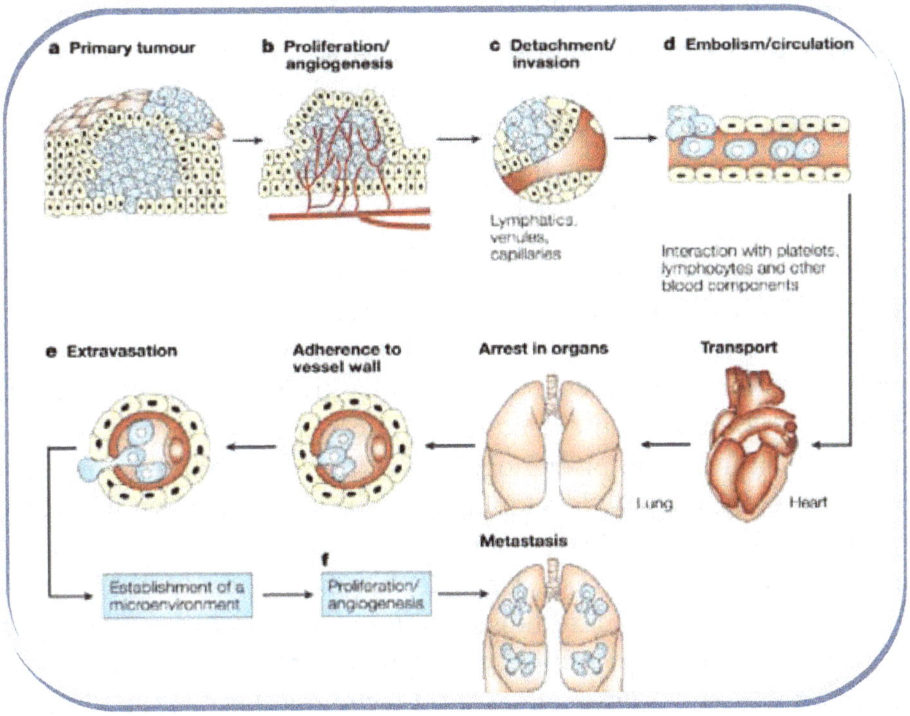

Figure 5. Tissue invasion is a multi-step process that requires the cancer cell to have developed many malignant traits, necessary for the formation of new distant colonies called metastases [27].

3. Mathematics of cancer

In comparison to biology, cell biology, and drug delivery analysis, mathematics has, to date, made comparatively very few contributions to this field of research. A statistical analysis of the PubMed platform list information (http://www.ncbi.nlm. nih.gov/PubMed/) showed that out of 1.5 million works that deal with cancer analysis, only 5% are associated with mathematical modeling. However, it is clear that mathematics could contribute significantly to areas of experimental cancer analysis since there is currently a wealth of experimental information which needs a systematic analysis.

Even in these conditions, in the last decade, mathematical modeling and machine simulation of cancer has multiplied dramatically (e.g., reviews like [30-35]. A broad range of strategies were developed, specializing in one or additional aspects of cancer. For example, genetic instability, natural selection or interactions of individual cell with each other or the environment have been modeled using methods of cellular automata and agent-based modeling.

These discrete methods have the disadvantage of being difficult to use when we deal with tumors of significant size. (see [36-38] for samples of cellular automata modeling and [39,40] for samples of agent-based modeling). In systems at larger scales, the neoplasic cell population is of the order of 10^6 or more, making these discrete methods unfitted. For these situations, the continuum methods provide the best approach. Early work, as well as [41-43], used ODE to model cancer as a uniform population and partial differential equation models restricted to spherical geometries. To assess the stability of spherical tumors to asymmetric perturbations and to characterize the degree of aggression [31, 44-48] use linear and weakly nonlinear analysis. The interactions of a growth with the microenvironment, like stress-induced limitations to growth, are studied in [30,49-54]. For the sake of simplicity, most of the modeling has considered single-phase (e.g., single cell species) tumors. To provide an elaborate account of growth non-uniformity, [50, 55, 56] have been developed a mixture of models.

The results of morphology instabilities on each avascular and vascular solid neoplasm growth have been recently studied using non-linear modeling. With the help of boundary integral methods, Cristini et al [47] performed the first absolutely nonlinear simulations of a time model of neoplasm growth within the avascular and vascular growth stages with arbitrary boundaries. The model from [47] has been extended in 3D by Li et al. [48] via adaptive boundary integral technique. The inclusion of angiogenesis and extratumoral environment has been performed by Zheng et al., [57]. By developing and coupling a level set implementation with a hybrid continuum-discrete growing model originally developed by Anderson & Chaplain [58]they found that low-nutrient (e.g., hypoxic) conditions could lead to morphological instability. Their work served as a building block for recent studies of the impact of therapy on neoplasm growth [59] and for studies of morphological instability and invasion [60-62]. Macklin & Lowengrub used a ghost cell/level set technique for evolving interfaces to check neoplasm growth in heterogeneous tissue and additional studied neoplasm growth as a function of the microenvironment [63]. Wise et al. [64] and Frieboes et al. [65] have developed a diffuse interface implementation of solid neoplasm growth for the study of the evolution of multiple neoplasm cell species, that was used in [65] to model the 3-D vascularised growth of malignant gliomas (brain tumors).

In biological systems, the fractal structure of area in which cells act and differentiate is important for their organization and emergence of the hierarchical network of multiple cross-interacting cells, sensitive to external and internal conditions. The biological phenomena occur within the area whose dimensions aren't represented solely by integers (1, 2, 3, etc.) of Euclidean space. Particularly, malignant tumors [53-56] grow in a space with non-integer dimension, i.e. fractal dimension. The analytical formulae describing the time-dependence of the temporal fractal dimension and scaling reproduce the expansion of the Flexner–Jobling rat's neoplasm in particular and growth of different rat's tumors generally. The results of some calculations indicated that the formula derived for the time-dependent temporal fractal dimension and the scaling factor describe the experimental data obtained by Schrek for the Brown-Pearce rabbit's neoplasm growth within the fractal time-space [3, 66-68].

In our assertion, fractal space(-time) consists in developing the consequences of the withdrawal of space(-time) differentiability's hypothesis and acquiring a fractal geometry, namely space(-time) becomes explicitly dependent on the observation scale [69].

On the other hand, of great use in our further reasonings will be the fact that in many biological systems it is possible to empirically demonstrate the presence of attractors that operate starting from different initial conditions (Ivancevic). Some of these attractors are points, some are closed curves, while the others have non–integer, fractal dimension and are termed "strange attractors" [70]. It has been proposed that a prerequisite for proper simulating tumor growth by computer is to establish whether typical tumor growth patterns are fractal. The fractal dimension of tumor outlines was empirically determined using the box-counting method [71]. In particular, fractal analysis of a breast carcinoma was performed using a morphometric method, which is the box-counting method applied to the mammogram as well as to the histological section of a breast carcinoma [72].

If tumor growth is chaotic, this could explain the unreliability of treatment and prediction of tumor evolution. More importantly, if chaos is established, this could be used to adjust strategies for fighting cancer. Treatment could include some form of chaos control and/or anti-control.

4. A few words about holography

"Although it generates a three-dimensional image, a hologram is most often recorded on a photographic plate or a flat piece of film. Moreover, producing a hologram does not imply, in the conventional sense, the recording of an image. To better understand this apparent paradox and, as a result, the way holography works, we have to begin with the main principles.

In conventional imaging techniques, e.g. photography, what is being recorded is merely the intensity distribution in the original scene. Thus, all information about the optical paths to different parts of the scene is lost.

The unique property of holography is the method of recording both the phase and the amplitude of the light waves from an object. Since all recording materials respond only to the intensity in the image, it is mandatory to convert the phase information into intensity variations. Holography accomplishes this by using coherent illumination and introducing, as shown in Fig. 6, a reference beam derived from the same source. The photographic film records the interference pattern produced by this beam and the light waves scattered by the object in cause.

Since the intensity at any given point in this pattern of interference also depends on the phase of the object wave, the resulting recording (the hologram) contains information on the phase as well as the amplitude of the object wave. If the hologram is illuminated once again with the original reference wave, as shown in Fig. 7, it reconstructs the original object wave.

An observer looking through the hologram sees a perfect three-dimensional image. This image exhibits all the effects of perspective, and depth of focus when photographed, that characterized the original object.

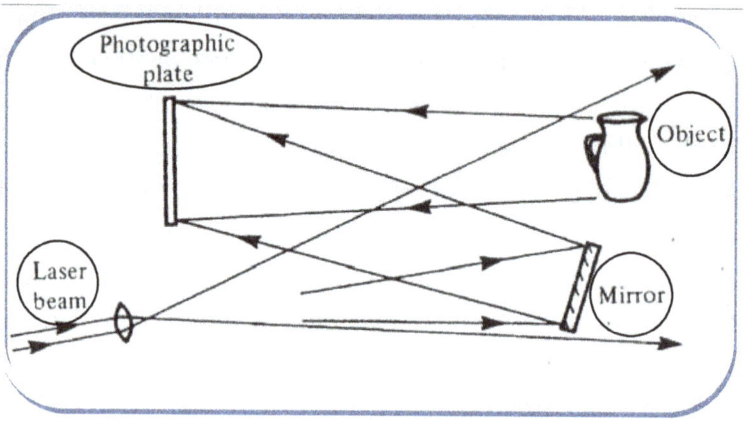

Figure 6. Hologram recording: the interference pattern produced by the reference wave and the object wave is recorded.

4.1. Early development

Gabor's historical experiment of holographic imaging [73] consisted in a transparency formed of opaque lines on a clear background which was illuminated with a collimated beam of monochromatic light, the interference pattern produced by the directly transmitted beam (the reference wave) and the light scattered by the lines on the transparency being recorded on a photographic plate. When the hologram (i.e. a positive transparency made from this photographic negative) was illuminated with the original collimated beam, it produced two diffracted waves, one which reconstructed an image of the object in its original location, and the other, with identical amplitude but an opposite phase, which formed a second, *conjugate* image.

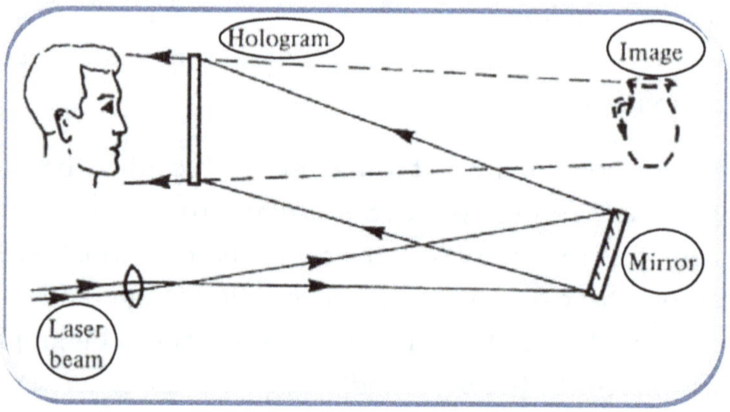

Figure 7. Image reconstruction: light diffracted by the hologram reconstructs the object wave.

An important flaw of this method of image reconstruction was the poor quality of the resulting image, due to the fact that it was degraded by the conjugate image that was superimposing on it as well as by the scattered light from the directly transmitted beam.

Leith and Upatnieks [74-76] found a solution to the above-mentioned problem developing a off-axis reference beam technique presented schematically in Figs. 6 and 7. They used a separate reference wave incident on the photographic plate at an appreciable angle to the object wave. As a result, when the hologram was illuminated with the original reference beam, the two images were separated by large enough angles from the directly transmitted beam, and from each other, thus ensuring that the images not overlap.

The improvement of the off-axis technique, and, in equal measure, the invention of the laser, which provided a powerful source of coherent light, resulted in a surge of activity in holography that led to several crucial applications.

4.2. The in-line hologram

Let us now look upon the optical system presented in Fig. 8 in which the object (a transparency containing small opaque details on a clear background) is illuminated by a collimated beam of monochromatic light along an axis normal to the photographic plate.

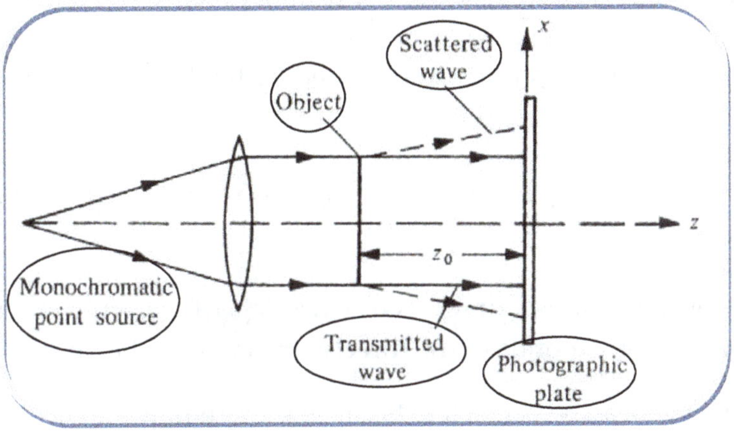

Figure 8. Optical system used to record an in-line hologram.

We can observe two components of the incident light. The first is the directly transmitted wave, which is a plane wave whose amplitude and phase do not vary across the photographic plate. Thus, its complex amplitude can be noted as a real constant r. The second one is a weak scattered wave whose complex amplitude at any point (x, y) on the photographic plate can be noted as $o(x, y)$, where $|o(x, y)| << r$.

From these it can be shown that the resulting complex amplitude is the sum of these two complex amplitudes, and because of that the intensity at this point is

$$I(x,y) = |r + o(x,y)|^2 = r^2 + |o(x,y)|^2 + ro(x,y) + ro^*(x,y) \tag{1}$$

where $o^*(x, y)$ is the complex conjugate of $o(x, y)$.

A 'positive' transparency (the hologram) is then made by contact printing from this recording. Therefore it can be assumed that this transparency is processed so that its amplitude transmittance (the ratio of the transmitted amplitude to that incident on it) can be written as

$$t = t_0 + \beta TI \tag{2}$$

where t_0 is a constant background transmittance, T is the exposure time and β is a parameter determined by the photographic material used and the processing conditions, the amplitude transmittance of the hologram is

$$t(x,y) = t_0 + \beta T[r^2 + |o(x,y)|^2 + ro(x,y) + ro^*(x,y)] \tag{3}$$

Then, the hologram is illuminated, as shown in Fig. 9, with the same collimated beam of monochromatic light employed to produce the original recording. Since the complex amplitude at any point in this beam is, aside from a constant factor, the same as that in the original reference beam, the complex amplitude transmitted by the hologram can be written as

$$u(x,y) = rt(x,y) = r(t_0 + \beta Tr^2) + \beta Tr|o(x,y)|^2 + \beta Tr^2 o(x,y) + \beta Tr^2 o^*(x,y) \tag{4}$$

The right-hand side of (4) contains four terms. The first, $r(t_0+\beta Tr^2)$, which represents a uniformly attenuated plane wave, corresponds to the directly transmitted beam.

The second, $\beta Tr\,|o(x, y)|^2$, can be neglected, because is extremely small, compared to the other terms.

The third term, $\beta Tr^2 o(x, y)$, is, except for a constant factor, identical with the complex amplitude of the scattered wave from the object and has the property of reconstructing an image of the object in its original position. Due to the fact that this image forms behind the hologram, and the reconstructed wave appears to diverge from it, it is a virtual image.

The fourth term, $\beta Tr^2 o^*(x, y)$, represents a wave similar to the object wave, but having an opposite curvature. This wave converges to form a real image (the conjugate image) at the same distance in front of the hologram.

With an in-line hologram, an observer viewing one image sees it superimposed on the out-of-focus twin image as well as a strong coherent background. Another problem is that the object must have a high average transmittance in order for the second term on the right-hand side of (4) to be negligible. Thus, it is possible to form images of fine opaque lines on a transparent background, but not *vice versa*. Finally, the hologram must be a 'positive' transparency. If the initial recording is used directly, β in (2) is negative, and the reconstructed image can be considered a photographic negative of the object.

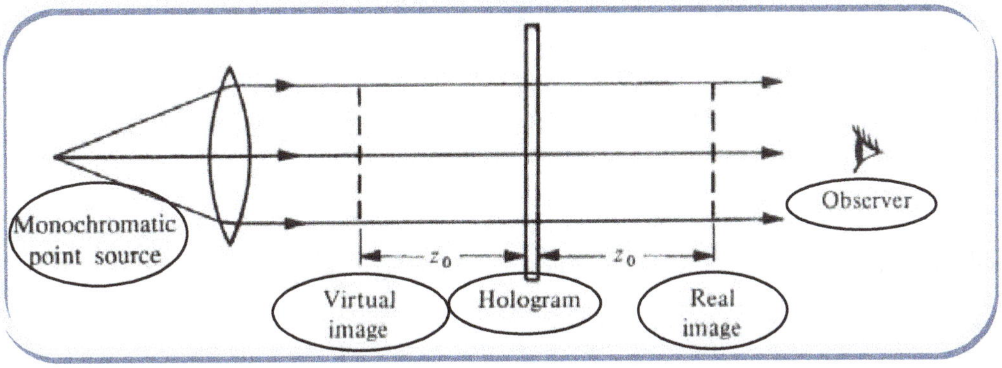

Figure 9. Optical system used to reconstruct the image with an in-line hologram, showing the formation of the twin images.

4.3. Off-axis holograms

In order to understand the formation of an image by an off-axis hologram, we must consider the recording arrangement shown in Fig. 10, in which (for simplicity) the reference beam is a collimated beam of uniform intensity, derived from the same source as the one used to illuminate the object.

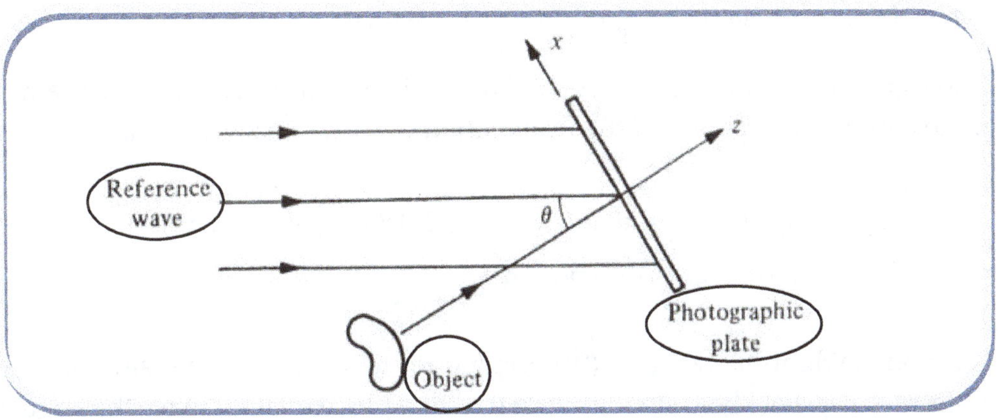

Figure 10. The off-axis hologram: `recording.

The complex amplitude at any point (x, y) on the photographic plate due to the reference beam can then be written as

$$r(x,y) = r\exp(i2\pi\xi x) \tag{5}$$

where ξ=(sinθ)/λ. Since only the phase of the reference beam varies across the photographic plat and because of the object beam, for which both the amplitude and phase vary, we can write the following:

$$o(x,y) = \left|o(x,y)\exp\left[-i\phi(x,y)\right]\right| \tag{6}$$

The resultant intensity is, therefore,

$$\begin{aligned} I(x,y) &= \left|r(x,y) + o(x,y)\right|^2 = \left|r(x,y)\right|^2 + \left|o(x,y)\right|^2 + \\ &+ r\left|o(x,y)\right|\exp\left[-i\phi(x,y)\right]\exp(-i2\pi\xi x) + r\left|o(x,y)\right|\exp\left[i\phi(x,y)\right]\exp(i2\pi\xi x) = \\ &= r^2 + \left|o(x,y)\right|^2 + 2r\left|o(x,y)\right|\cos[2\pi\xi x + \phi(x,y)] \end{aligned} \tag{7}$$

The amplitude and phase of the object wave are encoded as amplitude and phase modulation, respectively, of a set of interference fringes equivalent to a carrier with a spatial frequency of ξ.

If, as in (2), we assume that the amplitude transmittance of the processed photographic plate is a linear function of the intensity, the resultant amplitude transmittance of the hologram is

$$\begin{aligned} t(x,y) &= t_0' + \beta T\left|o(x,y)\right|^2 + \beta Tr\left|o(x,y)\right|\exp[-i\phi(x,y)]\exp[-i2\pi\xi x] + \\ &+ \beta Tr\left|o(x,y)\right|\exp[i\phi(x,y)]\exp[i2\pi\xi x] \end{aligned} \tag{8}$$

where $t_0' = t_0 + \beta Tr^2$ is a constant background transmittance.

When the hologram is illuminated for a second time with the original reference beam, as shown in Fig. 11, the complex amplitude of the transmitted wave can be written as

$$\begin{aligned} u(x,y) &= r(x,y)t(x,y) = t_0' r\exp(i2\pi\xi x) + \beta Tr\left|o(x,y)\right|^2 \exp(i2\pi\xi x) + \\ &+ \beta Tr^2 o(x,y) + \beta Tr^2 o^*(x,y)\exp(i4\pi\xi x) \end{aligned} \tag{9}$$

The first term on the right-hand side of (9) corresponds to the directly transmitted beam, while the second term generates a halo surrounding it, with approximately twice the angular spread of the object. The third term is identical to the original object wave, except for a constant factor βTr^2, and produces a virtual image of the object in its original position. The fourth term corresponds to the conjugate image which, in this case, is a real image. If the offset angle of the reference beam is taken large enough, the virtual image can be separated from the directly transmitted beam and the conjugate image.

In this setup, corresponding points on the real and virtual images are located at equal distances from the hologram, but on opposite sides of it. Because the depth of the real image is inverted, it is called a pseudoscopic image, as opposed to the normal, or orthoscopic, virtual image. It should also be mentioned that the phase of the reconstructed image is influenced only by the sign of β, always resulting in a "positive" image, even in the case of the hologram recording being a photographic negative.

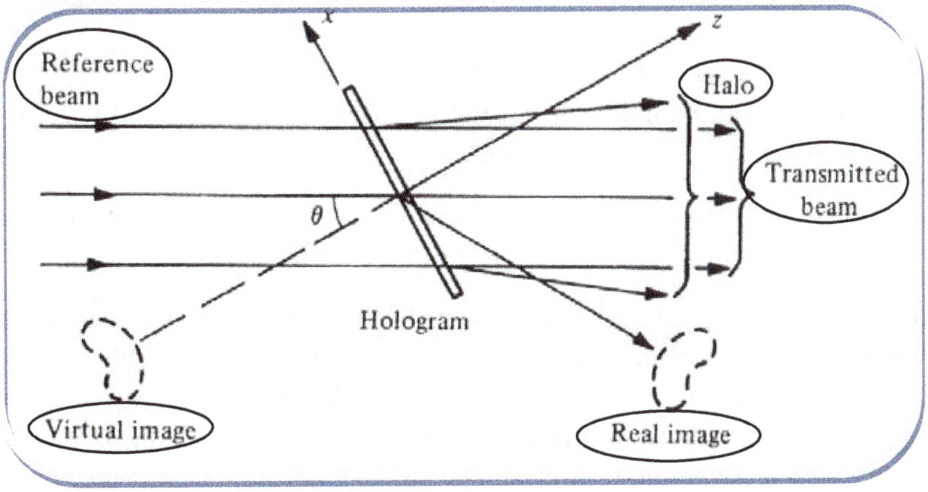

Figure 11. The off-axis hologram: image reconstruction.

4.4. Recording materials

Several recording materials have been used for holography [77]. Table 1 lists the principal characteristics of those that have been found most useful.

Material	Exposure J/m^2	Resolution mm^{-1}	Processing	Type	ηmax (diffraction efficiency)
Photographic	≈1.5	≈ 5000	Normal	Amplitude	0.06
			Bleach	Phase	0.60
DCG (dichromated gelatin)	10^2	10000	Wet	Phase	0.90
Photoresists	10^2	3000	Wet	Phase	0.30
Photopolymers	10-10^4	5000	Dry	Phase	0.90
PTP (photothermoplastics)	10^{-1}	500-1200	Dry	Phase	0.30
BSO ($Bi_{12}SiO_{20}$ photorefractive crystals)	10	10000	None	Phase	0.20

Table 1. Recording materials for holography

High-resolution photographic plates and films were the first materials used to record holograms. These are used widely even now, due to the fact that they exhibit relatively high sensitivity when compared to other hologram recording materials [78]. Moreover, they can be dye sensitized so that their spectral sensitivity matches the most commonly used laser wavelengths.

Combining the high sensitivity of photographic materials with the high diffraction efficiency, low scattering and high light-stability of DCG (dichromated gelatin) [79] was made possible by the silver-halide sensitized gelatin technique.

In positive photoresists, such as Shipley AZ-1350, the areas exposed to light become soluble and are washed away during development to produce a relief image [80].

Several organic materials can be activated by a photosensitizer to produce refractive index changes, because they suffer photopolymerization, when exposed to light [81]. A commercial photopolymer, coated on a polyester film base (DuPont OmniDex) that can be used to produce volume phase holograms with high diffraction efficiency is being currently produced [82].

Photothermoplastics (PTP) - a hologram can be recorded in a multilayer structure consisting of a glass or Mylar substrate coated with a thin, transparent, conducting layer of indium oxide, a photoconductor, and a thermoplastic [83,84].

When a photorefractive crystal is exposed to a spatially varying light pattern, electrons are liberated in the illuminated areas. These electrons migrate to adjacent dark regions, being trapped there. The spatially varying electric field produced by this space-charge pattern modulates the refractive index through the electro-optic effect, producing the equivalent of a phase grating. The space charge pattern can be removed by uniformly illuminating the crystal, after which another recording can take place [85,86].

It is essential to use coherent illumination for maximizing the visibility of the interference fringes formed by the object and reference beams, in the process of recording a hologram. In addition to being spatially coherent, the coherence length of the light must be much greater than the maximum value of the optical path difference between the object and the reference beams in the recording system. Lasers are, as a result, employed almost universally as light sources for recording holograms." (The text in quotation marks was reproduced from Hariharan P. [165]).

Consequently, to get a hologram, one needs a laser, which provides a powerful source of coherent light, and a 'recording material' which records the interference pattern produced by a reference beam and the light waves scattered by the object.

5. Tumor-associated ECM or tumor microenvironment, nonlinear medium with holographic properties

5.1. Extracellular matrix and tumor microenvironment

Within tissue, cells are surrounded by a meshwork of proteins and proteoglycans collectively called the extracellular matrix (ECM), which compartmentalizes tissues. The ECM is divided into two distinct layers:

i. the basement membrane, which is composed of sheet-like layers of ECM and lies under epithelial cells segregating tissues into functionally distinct regions;

ii. the interstitial matrix, which exists within intercellular space. The ECM serves multiple functions that are critical for embryonic development and wound repair. These functions include providing tissues with shape and flexibility and acting as a cushion to absorb external pressure. The ECM also serves as a base for cell anchorage, which mediates cell polarity, intercellular signaling, and assists in migration. The key to the ECM function lies in its unique composition and structure. The ECM is constructed in a specific pattern that is critical to its ability to carry out these functions and alterations in the expression level or arrangement of proteins within the ECM can be used to manipulate its function.

The most obvious function of the ECM is to provide structural support, shape, and stability for tissues. It does this by functioning as a base for cell anchorage. This base consists of three main structural components collagen, fibronectin, and elastic fibers, which bind to one another building a protein lattice upon which cells adhere.

Cell adherence to the ECM lattice provides cells support needed for cell migration. This is particularly important during embryonic development when cells are required to migrate into surrounding regions and differentiate into specific tissues [87]. A less obvious yet possibly more important function of the ECM in regards to tissue homeostasis and disease is its ability to mediate intracellular signaling. The ECM affects signaling through three main mechanisms:

i. cell – ECM interaction;

ii. regulation of the bioavailability of growth factors;

iii. the function of matricellular proteins. Cell attachment to the ECM via integrins induces signaling cascades that promote survival. Loss of cell-ECM contact can result in a form of apoptosis termed anoikis [88]. Anchorage-dependent survival is observed in most cells with the exception of red blood cells and inflammatory cells. However, tumor cells are often resistant to anoikis and can survive without a physical attachment to the ECM allowing them to successfully metastasize to distant tissues [89].

The ECM also affects cellular activity by serving as a reservoir for proteins required for proper tissue function and repair. This includes a plethora of growth factors and proteases. These pleiotropic molecules have been shown to robustly affect proliferation, survival and migration in numerous cell types. Once growth factors are secreted from cells, they often become embedded within the ECM and require ECM degradation by proteases such as elastase to release the active protein allowing it to interact with surrounding and transduce downstream signaling. The ability of the ECM to control the bioavailability of growth factors provides another means of regulating cellular activities and further explains how alterations in the makeup of the ECM as observed in diseases such as cancer affect cell response.

Matricellular proteins also reside in the ECM. They are a unique family of proteins that do not function as structural proteins but rather orchestrate the deposition of the ECM and mediate cell-cell and cell-ECM interactions. To do this, matricellular proteins interact directly with cell surface receptors, structural proteins, growth factors and proteases found within the ECM [90]. Their expression is found in every tissue begins early in development, persists throughout adulthood and is increased during tissue remodeling events. Matricellular proteins are critical regulators of many aspects of cell function including differentiation, survival, proliferation and migration making them necessary for proper tissue function. Not surprisingly, given their affect on cell-ECM mediated signaling pathways, matricellular proteins have been shown to strongly influence tumor growth.

For tumor cells to metastasize, the local ECM must be remodeled to create an environment conducive to tumor survival and progression. This includes altering the architecture and composition of the tumor-associated ECM or tumor microenvironment (TME) to facilitate tumor cell dissemination [91]. Changes in ECM architecture are primarily carried out by enzymes such as MMPs which assist in remodeling of the TME by degrading structural proteins such as collagen and fibronectin allowing tumor cells to freely navigate through the surrounding ECM. MMPs and other proteases assist in destruction of the first barrier tumor cells face to successful metastasis, the basement membrane. They degrade the underlying basement membrane allowing tumor cells to escape the primary tumor and invade into surrounding non-neoplasic tissues. MMPs continue to breakdown barriers in the surrounding ECM clearing a path to blood vessels where tumor cells will intravasate into the circulatory system and seed secondary tumors [92]. Destruction of the ECM by proteases also promotes tumor progression by facilitating the release of angiogenic and mitogenic factors bound within the ECM [93]. In a surprising unexpected twist, studies revealed that the breakdown of ECM proteins by MMPs was more complex than anticipated. In fact it is a highly organized process which results in the generation of both protumor and antitumor cleavage products [94].

Presence of the ECM is required for cellular survival therefore increased degradation of the ECM within the TME must be balanced by an increase in ECM synthesis. The development of a tumor, much like a wound, provokes a robust inflammatory response causing an influx of mast cells, macrophages and neutrophiles into the TME [95].

We may summarize that the extracellular matrix generates signaling cues that regulate cell behavior and orchestrate functions of cells in tissue formation and homeostasis. Microenvironmental signaling, a process that determines cell shape, motility, growth, survival and differentiation is highly influenced by the ECM properties: composition, three-dimensional organization and proteolytic remodeling. Recent studies have shown that misregulation of cell–ECM interactions can contribute to many diseases, including developmental, immune, haemostasis, degenerative and malignant disorders.

Consequently, the structure and the behavior of the tumor-associated ECM allows us to think of it as a non-differential medium, and as will be shown below, a medium which holds the properties of a hologram (capacity to memorize, interference abilities) and may become a source of forces. In other words, ECM and TME are very suitable candidates for a 'recording material'.

5.2. Tumor-associated ECM as a non-differential fractal medium

We can simplify the dynamics of a biological system supposing that the motions on ECM take place on continuous but non-differentiable curves, i.e. fractal curves (for example, the Peano curve, the Koch curve or the Weierstrass curve [69,96,97].

Once this hypothesis has been accepted, some consequences of non-differentiability by SRT are evident [69,96]: i) the physical quantities that are used in describing the biological system dynamics are fractal functions, i.e. functions dependent both on spatial coordinates and time as well as on the scale resolution, $\delta t/\tau$ (identified here with dt/τ by means of the substitution principle [69,96]. We mention that in the standard biophysics, the physical quantities describing the dynamics of a biological system are continuous, but differentiable functions depending only on spatial coordinates and time; ii) the dynamics of the biological systems are given by the fractal operator $\hat{d}/dt$ [98]:

$$\frac{\hat{d}}{dt}=\frac{\partial}{\partial t}+\hat{V}\cdot\nabla-i\frac{\lambda^2}{\tau}\left(\frac{dt}{\tau}\right)^{\left(\frac{2}{D_F}\right)-1}\Delta \tag{10}$$

where

$$\hat{\mathbf{V}}=\mathbf{V}_D-i\mathbf{V}_F \tag{11}$$

is the complex velocity, V_D is the differentiable and resolution scale independent velocity, V_F is the non-differentiable and resolution scale dependent velocity, $\hat{V}\ \nabla$ is the convective term,

$$\frac{\lambda^2}{\tau}\left(\frac{dt}{\tau}\right)^{\left(\frac{2}{D_F}\right)-1}\Delta=\frac{\lambda^2}{\tau}\left(\frac{dt}{\tau}\right)^{\left(\frac{2}{D_F}\right)-1}\left(\frac{\partial^2}{\partial x^2}+\frac{\partial^2}{\partial y^2}+\frac{\partial^2}{\partial z^2}\right) \tag{12}$$

is the dissipative term, D_F is the fractal dimension of the movement curve, λ is the space scale, τ is the time scale and λ^2/τ is a coefficient specific to the fractal – non - fractal transition. For D_F any definition can be used (the Hausdorff – Besikovici fractal dimension, the Kolmogorov fractal dimension, etc. [97], but once such definition is accepted for D_F, it has to remain constant over the entire analysis of the complex fluid dynamics. In a particular case, for motions on Peano curves, $D_F = 2$ [97] of the complex fluid entities, the fractal operator (1) is reduced to Nottale's operator $(\hat{d}/dt)_N$

$$\frac{\hat{d}}{dt}=\frac{\partial}{\partial t}+\hat{\mathbf{V}}\cdot\nabla-iD_N\Delta$$

where $D_N = \lambda^2/\tau$ is the Nottale's coefficient associated to the fractal-non-fractal transition.

Applying the fractal operator (10) to the complex velocity (11) and accepting the principle of scale covariance [69,96] in the form:

$$\frac{\hat{d}\hat{\mathbf{V}}}{dt} = -\nabla U \tag{13}$$

we obtain the motion equation:

$$\frac{\hat{d}\hat{\mathbf{V}}}{dt} = \frac{\partial \hat{\mathbf{V}}}{\partial t} + \left(\hat{\mathbf{V}} \cdot \nabla\right)\hat{\mathbf{V}} - i\frac{\lambda^2}{\tau}\left(\frac{dt}{\tau}\right)^{\left(\frac{2}{D_F}\right)-1} \Delta\hat{\mathbf{V}} = -\nabla U \tag{14}$$

where U is an external scalar potential. Equation (14) is a Navier – Stokes type equation. It means that at any point of a fractal path, the local acceleration term, $\partial_t \hat{V}$, the non-linearly (convective) term, $(\hat{V}\ \nabla)\ \hat{V}$, the dissipative term, $(\lambda^2/\tau)(dt/\tau)^{\left(\frac{2}{D_F}\right)-1}\Delta\hat{V}$, and the external free term ∇U make their balance. Therefore, the biological fluid is assimilated to a "rheological" fractal fluid, whose dynamics are described by the complex velocities field, $\hat{V}$, and by the imaginary viscosity type coefficient, $i(\lambda^2/\tau)(dt/\tau)^{\left(\frac{2}{D_F}\right)-1}$. The "rheology" of the fractal fluid can provide hysteretic properties to the biological fluid (the fractal fluid has a hysteresis cycle, memory, etc. [98-100].

For irrotational motions of the biological system entities

$$\begin{aligned} &\nabla \times \hat{\mathbf{V}} = 0, && a \\ &\nabla \times \hat{\mathbf{V}}_D = 0, && b \\ &\nabla \times \hat{\mathbf{V}}_F = 0 && c \end{aligned} \tag{15}$$

we can choose $\hat{V}$ of the form

$$\hat{\mathbf{V}} = -i\frac{\lambda^2}{\tau}\left(\frac{dt}{\tau}\right)^{\left(\frac{2}{D_F}\right)-1} \nabla \ln \psi \tag{16}$$

where $\phi \equiv \ln\psi$ is the velocity scalar potential. By substituting (16) in (14) and using the method described in [98-100], it results

$$\frac{\hat{d}\hat{\mathbf{V}}}{dt} = -i\frac{\lambda^2}{\tau}\left(\frac{dt}{\tau}\right)^{\left(\frac{2}{D_F}\right)-1} \nabla\left[\frac{\partial \ln\psi}{\partial t} - i\frac{\lambda^2}{\tau}\left(\frac{dt}{\tau}\right)^{\left(\frac{2}{D_F}\right)-1}\frac{\nabla\psi}{\psi} + U\right] = 0 \tag{17}$$

This equation can be integrated in a universal way and yields

$$\frac{\lambda^4}{\tau^2}\left(\frac{dt}{\tau}\right)^{\left(\frac{4}{D_F}\right)-2}\Delta\psi + i\frac{\lambda^2}{\tau}\left(\frac{dt}{\tau}\right)^{\left(\frac{2}{D_F}\right)-1}\frac{\partial\psi}{\partial t} - \frac{U}{2}\psi = 0 \tag{18}$$

up to an arbitrary phase factor which may be set to zero by an appropriate selection of the phase of ψ. Relation (18) is a Schrödinger type equation. For motions on Peano curves, $D_F = 2$ [97] at Compton scale, which implies $\lambda^2/\tau = \hbar/2m_0$ [69,96], with $\hbar$ the reduced Plank constant and m_0 the rest mass of the biological entities, the relation (18) becomes the standard Schrödinger equation:

$$\frac{\hbar^2}{2m_0}\Delta\psi + i\hbar\frac{\partial\psi}{\partial t} - \frac{U}{2m_0}\psi = 0$$

If $\psi = \sqrt{\rho}\, e^{iS}$, with $\sqrt{\rho}$ the amplitude and S the phase of ψ, the complex velocity field (16) takes the explicit form:

$$\begin{aligned}
\hat{\mathbf{V}} &= \frac{\lambda^2}{\tau}\left(\frac{dt}{\tau}\right)^{\left(\frac{2}{D_F}\right)-1}\nabla S - i\frac{\lambda^2}{2\tau}\left(\frac{dt}{\tau}\right)^{\left(\frac{2}{D_F}\right)-1}\nabla \ln\rho \quad &a\\
\mathbf{V}_D &= \frac{\lambda^2}{\tau}\left(\frac{dt}{\tau}\right)^{\left(\frac{2}{D_F}\right)-1}\nabla S \quad &b\\
\mathbf{V}_F &= \frac{\lambda^2}{2\tau}\left(\frac{dt}{\tau}\right)^{\left(\frac{2}{D_F}\right)-1}\nabla \ln\rho \quad &c
\end{aligned} \tag{19}$$

By substituting (19a-c) in (14) and separating the real and the imaginary parts, up to an arbitrary phase factor which may be set to zero by appropriate selection of the phase of ψ, we obtain:

$$\frac{\partial \mathbf{V}_D}{\partial t} + (\mathbf{V}_D\cdot\nabla)\mathbf{V}_D = -\nabla(Q+U) \tag{20}$$

$$\frac{\partial\rho}{\partial t} + \nabla\cdot(\rho\mathbf{V}_D) = 0 \tag{21}$$

with Q the specific fractal potential

$$Q = -2\frac{\lambda^4}{\tau^2}\left(\frac{dt}{\tau}\right)^{\left(\frac{4}{D_F}\right)-2}\frac{\Delta\sqrt{\rho}}{\sqrt{\rho}} = -\frac{\mathbf{V}_F^2}{2} - \frac{\lambda^2}{\tau}\left(\frac{dt}{\tau}\right)^{\left(\frac{2}{D_F}\right)-1}\nabla\cdot\mathbf{V}_F \tag{22}$$

Equation (20) represents the specific momentum conservation law, while equation (21) represents the states density conservation law. By means of the fractal velocity, V_F, the specific fractal potential Q is a measure of non-differentiability of the biological entities trajectories, *i.e.* of their chaoticity. The equations (20)-(22) define the fractal hydrodynamics model (FHM). In such a context, the biological system can be considered a fractal fluid.

Thus, it can be concluded that: i)Any entity of the biological system is in a permanent interaction with the fractal medium by means of the specific fractal potential; ii) The fractal medium is identified with a non-relativistic fractal fluid described by equations (20)-(22); iii) For motions on Peano curves at Compton scale [69,96,97], the FHM reduces to a quantum hydrodynamic model (QHM). Indeed, according to our previous considerations the relations (19a-c) become

$$\hat{\mathbf{V}} = \frac{\hbar}{m_0}\nabla S - i\frac{\hbar}{2m_0}\nabla \ln \rho$$

$$\mathbf{V}_D = \frac{\hbar}{m_0}\nabla S$$

$$\mathbf{V}_F = \frac{\hbar}{m_0}\nabla \ln \rho$$

in order that the momentum and density conservation laws are given by (20) and (21), respectively, with V_D and V_F previously defined, and the specific fractal potential by the expression

$$Q = -\frac{\hbar^2}{2m_0}\frac{\nabla^2\sqrt{\rho}}{\sqrt{\rho}} = -\frac{\mathbf{V}_F^2}{2} - \frac{\hbar}{2}\nabla\cdot\mathbf{V}_F$$

Moreover the fractal medium is assimilated to Bohm subquantum level [96]; iv) The fractal velocity V_F cannot be regarded as actual mechanical motion; it contributes to the transfer of the specific momentum and the concentration of energy. This fact can easily be deduced from the absence of V_F in the states density conservation law, and from its role in the variational principle. Any interpretation of Q should take into account the "self" or internal nature of the specific momentum transfer. While the energy is stored in the form of mass motion and potential energy (as it is classically), a part of it is available elsewhere and only the total is conserved. Reversibility and the existence of eigenstates is ensured by the conservation of energy and specific momentum, but this also means that a Brownian motion [97] form of interaction with an external medium is denied; v) For Peano curves motions [96,97], at spatial scales higher than the dimension of the boundary layer and at temporal scales higher than the oscillation periods of the pulsating velocities which overlaps the average velocity of the biological fluid motions (for details see [101-103], the FHM reduces to the standard hydrodynamical model [104]; vi) Since the position vector of the biological system entity is assimilated with a stochastic Wiener type process [96,97], ψ is not only the scalar potential of a complex velocity (through $\phi \equiv \ln\psi$) in the frame of FHM, but represents also the states density (through ψ^2) in the frame of a Schrödinger type model. Thus it can be seen that the formalism of the FHM and the one of Schrödinger type are equivalent. In addition, the chaoticity, either by means of turbulence in the fractal hydrodynamics approach, or by means of stochasticization in the Schrödinger type approach, is generated only by the non-differentiability of the movement trajectories in a fractal space; vii) In the standard model (Landau's scenario [104]) the Fourier spectrum is always discrete and cannot approximate a continuum spectrum that in case of a large number of frequencies will generate an unlimited number of spectral components as a result of their beats which appear due to the presence of nonlinearities in the biological fluid. Still, taking into account the standard model, the flow can never be exactly chaotic because, in case of multiple periodic functions, correlations tend to be null, although having an oscillating character. As a result, the transition towards chaotic behavior can be described by Landau's scenario only in a biological system with an infinite number of degrees of freedom. In our case, when $\delta t/\tau \rightarrow 0$ for $D_F \neq 2$ the physical quantities that describe the dynamics of the biological system are no longer defined. So, in this approximation, a simulation of a system with an infinite number of degrees of freedom is used. Moreover, the possibility of the dynamic states generation should be noted, which is characterized by windows of regular oscillations interrupted by chaotic bursts, the transition between the two states being spontaneous, unpredictable and independent of any of the control parameters variation (turbulence through intermittency); viii) The fractal medium and in particular the subquantum level has some computational properties: viii1) bistability, which implies the existence of its fractality and in particular, for motions on Peano curves at Compton scales, of the quantum bit. And from here, the entire fractal logics and in particular the quantum one; viii2) the self-replication, which implies the existence of some specific self-copying mechanisms; viii3) memory, which implies hysteresis type mechanisms; viii4) self-similarity, which implies the holographic type behavior; viii5) polarization, which implies mechanisms of changing the "computational state" from a given to a desired one; viii6) depositing and transmitting the information etc. For details see [105].

6. Tumor self-seeding by CTC and hypoxia support the idea of complete holography

6.1. The self-seeding hypothesis of tumor growth

The unfolding of cancer cells from their original sites to different ones within the body, i.e metastasis, has, for many years, been regarded as a unidirectional journey. However some researchers conjointly consider that metastatic cancer cells can increase primary tumor growth, this fact being crucial for the planning and type of the cancer treatment.

The concept of growing self-metastasis, or tumor "self-seeding," was first introduced at Memorial Sloan-Kettering Cancer Center, in a range of studies conducted by Drs. Joan Massagué, head of the Metastasis research facility, and Larry Norton, deputy physician-in-chief of the center's breast cancer programs. In the studies conducted on mice, Dr. Massagué discovered that breast tumors express genes related to metastasis were growing quicker than tumors that didn't express these genes, even if the genes had no apparent role in increased cellular division or decreased cell death (Fig. 12). These results did not fit within the standard tumor growth theories. In 2006, the researchers theorized that cells that become independent from a tumor and colonize distant tissues may also return home to the microenvironment within which they initial developed via the cardiovascular system [106]. They tested their hypothesis in a mouse model of cancer and revealed their findings in 2009 in Cell [107].

In one particular experiment, they selected a non-metastatic breast cancer cell line and an isolated set of daughter cells from that line that had gained the ability over time to metastasize to the lungs. Consequently, they implanted the parent cells in one mammary gland and the metastatic daughter cells in the opposite gland to serve as "donor tumors". They noticed that the daughter cells migrated to the lungs and to the tumors that were being formed by the parent cells in the opposite organ, accounting for 5 to 30% of the size of the parent tumors. Also, it was obvious that parent tumors seeded by daughter cells grew quicker than parent tumors that were implanted without daughter cells within the opposite gland.

This specific seeding behavior with daughter cells that spread to the bones and brain was noticed in the studies of colon cancer and skin cancer cell lines, but not when non-metastatic daughter cells were transplanted.

Furthermore, in various related follow-up laboratory experiments, the researchers demonstrated that cells from primary tumors can attract circulating metastatic tumor cells, and found several proteins that probably encourage this migration. They also found that the "come-back" metastatic cells mainly influenced primary tumor growth through the release of proteins that modify the tumor microenvironment, as well as blood vessels and immune cells.

Their hypothesis started to obtain support from different researchers at the United Nations agency. In 2009, Dr. Philip Hahnfeldt and his colleagues published the results of computer modeling studies conceived to search at the intersection of two biological phenomena found in tumors [108]. One of these phenomena consists in a small population of cancer cells acting like stem cells; thus, they could possess the ability to reproduce an infinite number of times, and also to generate secondary cancer cells that, with time, lose the ability to divide. The second phenomena is that tumor growth is restricted by the available space for growing. Normally, healthy cells are spatially separated by a space that is not available in the tumor, due to the fact that cancer cells grow tightly in a dense mass till all the available has been occupied and the cellular division stops. At the periphery of the tumor, where the normal tissues density decreases, cancer cells continue to multiply and expand, increasing the size of the tumor.

Their models showed a crucial relation between cell migration, cell death, and tumor growth. When the offspring of a cancer stem cell within the model did not migrate or die spontaneously, tumor growth remained constant at around 110 cells. On the other hand, a high death rate among the non-stem cell progeny combined with a high cell migration rate produced the biggest tumors within the shortest amount of your time, to virtually 100,000 cells in over three years.

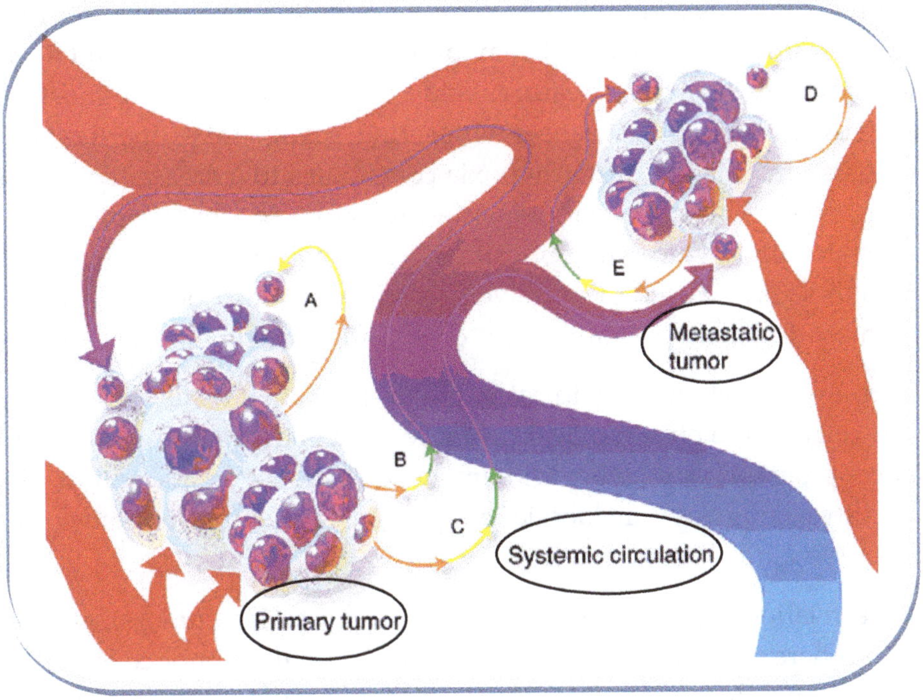

Figure 12. In the self-seeding concept of cancer growth and metastasis, a mobile tumor cell can take one of five different pathways in the body. *A* – evade and return to the primary tumor, using only the close ECM and not the systemic circulation; *B* – escape into the systemic circulation and then return to the original tumor; *C* – migrate through the systemic circulation and grow a metastatic tumor elsewhere in the organism; *D* – evade and return to the metastatic tumor, not using the systemic circulation; *E* –escape and return to the metastatic tumor through the systemic circulation.

This theoretical phenomena – accelerated growth jump-started by a high rate of growth death – has potential implications for the clinical treatment of cancer. Traditional cytotoxic therapy medicine kill massive numbers of speedily dividing cancer cells, however might not have an effect on cancer stem cells in each tumor type.

In the light of the above, we think of the CTC returning to the initial tumor site and fueling the primary tumor growth or even grow a new tumor as a particular case of complete holography (*i.e.* a hologram which does not represent only the virtual object's image, but becomes the very object - which we believe, is a characteristic of the living organisms).

6.2. Hypoxia and cancer

Vascularized tissues is the trigger factor for a large number of cellular processes, combined with an adaptive response. [109,110]. Following the drop in oxygen supplies, cells start to adapt to the less favorable environment and to initiate a vascularization| process in order for them to raise the local oxygen supply. In the center of the hypoxic response is the angiogenic shift, with production of potent angiogenic factors such as vascular endothelial growth factor (VEGF). Hypoxia, which is present in many solid malignancies, means that oxygen level in tumors corresponds to around 1.5% [109]. This is caused on one hand by the result of the abnormal vascularisation in tumors, that is short in supply oxygen to the sometimes rapidly expanding malignant lesion and on the other hand, the existence of areas with acute lack of oxygen, resulting in necroses and cell death on a large scale. In breast cancer these forms are usually related to clinically aggressive behavior. Markers for hypoxia like HIF-1a have been connected to extremely malignant features and could be relevant prognostic markers for distinguishing subgroups of breast cancer with certain malignant properties [111,112]. There is a current discussion whether or not hypoxia contributes to increase the aggressiveness of tumors or if aggressive tumors have more widespread hypoxia, but, apparently, one explanation doesn't essentially exclude the opposite. Recent analysis has shown that 'the hypoxia response' in tumors may be used to conceive new treatment methods [113]. Emerging cancer therapies will most definitely put more focus on specific targeting of hypoxic processes.

Human cancers are characterized by intratumoral hypoxia that results from the proliferation of deregulated cell and the physiological responses that is triggered by it have impact on all aspects of cancer progression, together with im mortalization, transformation, differentiation, genetic instability, ontogeny, metabolic adaptation, autocrine protein communication, invasion, metastasis, and resistance to therapy.

We assume the relationship between hypoxia and aggressive tumors may be due to the presence of the coherent wave laser with oxygen of metastatic tumor cells in the area, where the produced oxygen gradients lead to oxygen consumption. It has been already shown that laser photocoagulation is effective in the treatment of diabetic retinopathy, in a series of major studies [114-117]. The oxygen-consumption may be based on a multilayer solution to Fick's law of diffusion, yet the essence is that the oxygen consumption is greatest where the oxygen gradient changes most rapidly [118-120].

All the above considerations and hypoxia's impact on all critical aspects of cancer progression support the idea that, the metastatic tumor cells moving through the systemic circulation (and not necessarily in there), may be considered a travelling wave chemically pumped type laser with oxygen.

7. Basic model

7.1. The PDE cancer-invasion model

We consider and present in what follows in extenso, the basic mathematical model of growth of a generic solid tumor, which is assumed just been vascularised, i.e. a blood supply has been established. Let us focus on four key variables involved in tumor cell invasion, in order to produce a minimal model, namely tumor cell density (denoted by n), matrix-degradative enzymes (MDE) concentration (denoted by m), the complex mixture of macromolecules from the extracellular material's (MM) concentration (denoted by f) and the oxygen concentration (denoted by c). Each of the four variables (n, m, f, c) is a function of the spatial variable $\mathbf{x}$ and time t. Firstly, we have to define a system of coupled non-linear partial differential equations to model tumor invasion of surrounding tissue.

We make the assumption that the ECM is a mixture of MM (*e.g.* collagen, fibronectin, laminin and vitronectin) only and not any other cells. Most of the MM of the ECM which are important for cell adhesion, spreading and motility are *fixed* or *bound* to the surrounding tissue. MDEs are important at many stages of tumor growth, invasion and metastasis, and they interact with inhibitors, growth factors and tumor cells in a very complex way. Yet it is widely accepted that tumor cells produce MDEs which locally degrade the ECM. As well as creating space into which tumor cells may be transported by simple diffusion (random motility), we can assume that this also results in a gradient of these bound cell-adhesion molecules, such as fibronectin. As a result, while the ECM may be a barrier to normal cell movement, it also represents a substrate to which cells may adhere and move upon. The presence of a minimum of ECM elements is a requirement for the growth and survival of most mammalian cells, and indeed these cell will migrate up a gradient of bound (*i.e.* non-diffusible) cell-adhesion molecules in the *in vitro* cultures [121-126].

We can define haptotaxis as a directed migratory response of the cells to gradients of fixed or bound chemicals (i.e. non-diffusible chemicals). While studies have not yet clearly shown haptotaxis occurs in an *in vivo* situation, given the structure of human tissue, it is not without reason to assume that haptotaxis is a major component of directed movement in tumor cell invasion. Indeed, there has been much recent effort to characterize such directed movement [125-127]. We therefore will treat this directed movement of tumor cells in this model as haptotaxis, i.e. a response to gradients of bound MM such as fibronectin. To incorporate this response in the mathematical model, we take the haptotactic flux to be $\mathbf{J}_{\text{hapto}} = \chi\, n\, \nabla f$, where $\chi > 0$ is the (constant) haptotactic coefficient.

As we stated early, the only other contribution to tumor cell motility in this model is the random motion. To describe the random motility of the tumor cells, we assume a flux of the form $\mathbf{J}_{\text{rand}} = -D_n \nabla n$, where D_n is the constant random motility coefficient.

We only model the tumor cell migration at this level, as all other tumour cell processes, such as proliferation, adhesion and death will be treated at a single cell level within the hybrid discrete-continuum model. The conservation equation for the tumour cell density n can therefore be written as

$$\frac{\partial n}{\partial t} + \nabla\left(\mathbf{J}_{rand} + \mathbf{J}_{hapto}\right) = 0$$

and hence the partial differential equation governing tumor cell motion (in the absence of cell proliferation) is

$$\frac{\partial n}{\partial t} = D_n \nabla^2 n - \chi \nabla\left(n \nabla f\right) \tag{23}$$

The ECM is known to contain many MM, including fibronectin, laminin and collagen, which can be degraded by MDEs [128,129]. We assume that the MDEs degrade ECM upon contact and hence the degradation process is modeled by the following simple equation

$$\frac{\partial f}{\partial t} = -\delta mf \tag{24}$$

where δ is a positive constant.

Active MDEs are produced (or activated) by the tumor cells, diffuse throughout the tissue and undergo some form of decay (either passive or active). The equation governing the evolution of MDE concentration is therefore given by

$$\frac{\partial m}{\partial t} = D_m \nabla^2 m + g(n,m) - h(n,m,f) \tag{25}$$

where D_m is a positive constant, the MDE diffusion coefficient, g is a function modeling the production of active MDEs by the tumor cells and h is a function modeling the MDE decay. For simplicity we assume that there is a linear relationship between the density of tumor cells and the level of active MDEs in the surrounding tissues (not taking into consideration the amount of enzyme precursors secreted and the presence of endogenous inhibitors) and so these functions will be $g = \mu\, n$ (MDE production by the tumor cells) and $h = \lambda\, m$ (natural decay), respectively.

The fact solid tumors need oxygen to grow and invade is a well-known one. Oxygen is assumed to diffuse into the MM, decay naturally and be consumed by the tumor. We assume that oxygen production is proportional to the MM density. This is a crude way of modeling an angiogenic oxygen supply for a more appropriate way of modeling the angiogenic network. The oxygen equation then has the form,

$$\frac{\partial c}{\partial t} = D_c \nabla^2 c + \beta f - \gamma n - \alpha c \tag{26}$$

where D_c, β, γ, α are positive constants representing the oxygen diffusion coefficient, production, uptake and natural decay rates, respectively.

The complete system of equations describing the interactions of the tumor cells, MM, MDEs and oxygen as detailed above, is

$$\begin{aligned}
\frac{\partial n}{\partial t} &= \overbrace{D_n \nabla^2 n}^{\text{random motility}} - \overbrace{\chi \nabla (n \nabla f)}^{\text{haptotaxis}} && \text{a} \\
\frac{\partial f}{\partial t} &= -\overbrace{\delta mf}^{\text{degradation}} && \text{b} \\
\frac{\partial m}{\partial t} &= \overbrace{D_m \nabla^2 m}^{\text{diffusion}} + \overbrace{\mu n}^{\text{production}} - \overbrace{\lambda m}^{\text{decay}} && \text{c} \\
\frac{\partial c}{\partial t} &= \overbrace{D_c \nabla^2 c}^{\text{diffusion}} + \overbrace{\beta f}^{\text{production}} - \overbrace{\gamma n}^{\text{uptake}} - \overbrace{\alpha c}^{\text{decay}} && \text{d}
\end{aligned} \tag{27}$$

where D_n, D_m and D_c are the tumor cell, MDE and oxygen diffusion coefficients, respectively, χ is the haptotaxis coefficient and δ, μ, λ, β, γ and α are positive constants. We should also note that cell–matrix adhesion is modeled here by the use of haptotaxis in the cell equation, i.e. directed movement up gradients of MM. Therefore, χ maybe considered as relating to the strength of the cell–matrix adhesion.

7.2. The PDE cancer-invasion model via scale relativity theory

The presence of the fractal medium implies the substitution of the standard derivative d/dt with the fractal operator (10). Then the system (27a-d) becomes

$$\begin{aligned}
&\frac{\hat{d}n}{dt}=\frac{\partial n}{\partial t}+\left(\hat{\mathbf{V}}\cdot\nabla\right)n-i\frac{\lambda^2}{\tau}\left(\frac{dt}{\tau}\right)^{(2/D_F)-1}\Delta n=D_n\nabla^2 n-\chi\nabla\left(n\nabla f\right) && \text{a}\\
&\frac{\hat{d}f}{dt}=\frac{\partial f}{\partial t}+\left(\hat{\mathbf{V}}\cdot\nabla\right)f-i\frac{\lambda^2}{\tau}\left(\frac{dt}{\tau}\right)^{(2/D_F)-1}\Delta f=-\delta mf && \text{b}\\
&\frac{\hat{d}m}{dt}=\frac{\partial m}{\partial t}+\left(\hat{\mathbf{V}}\cdot\nabla\right)m-i\frac{\lambda^2}{\tau}\left(\frac{dt}{\tau}\right)^{(2/D_F)-1}\Delta m=D_m\nabla^2 m+\mu n-\lambda m && \text{c}\\
&\frac{\hat{d}c}{dt}=\frac{\partial c}{\partial t}+\left(\hat{\mathbf{V}}\cdot\nabla\right)c-i\frac{\lambda^2}{\tau}\left(\frac{dt}{\tau}\right)^{(2/D_F)-1}\Delta c=D_c\nabla^2 c+\beta f-\gamma n-\alpha c && \text{d}
\end{aligned}\tag{28}$$

or more explicitly, by separating the scales of interaction, for the differentiable scale

$$\begin{aligned}
&\frac{\partial n}{\partial t}+\left(\hat{\mathbf{V}}_D\cdot\nabla\right)n=D_n\nabla^2 n-\chi\nabla\left(n\nabla f\right) && \text{a}\\
&\frac{\partial f}{\partial t}+\left(\hat{\mathbf{V}}_D\cdot\nabla\right)f=-\delta mf && \text{b}\\
&\frac{\partial m}{\partial t}+\left(\hat{\mathbf{V}}_D\cdot\nabla\right)m=D_m\nabla^2 m+\mu n-\lambda m && \text{c}\\
&\frac{\partial c}{\partial t}+\left(\hat{\mathbf{V}}_D\cdot\nabla\right)c=D_c\nabla^2 c+\beta f-\gamma n-\alpha c && \text{d}
\end{aligned}\tag{29}$$

and for the fractal scale

$$\begin{aligned}
&\left(\hat{\mathbf{V}}_F\cdot\nabla\right)n=\frac{\lambda^2}{\tau}\left(\frac{dt}{\tau}\right)^{(2/D_F)-1}\Delta n && \text{a}\\
&\left(\hat{\mathbf{V}}_F\cdot\nabla\right)f=\frac{\lambda^2}{\tau}\left(\frac{dt}{\tau}\right)^{(2/D_F)-1}\Delta f && \text{b}\\
&\left(\hat{\mathbf{V}}_F\cdot\nabla\right)m=\frac{\lambda^2}{\tau}\left(\frac{dt}{\tau}\right)^{(2/D_F)-1}\Delta m && \text{c}\\
&\left(\hat{\mathbf{V}}_F\cdot\nabla\right)c=\frac{\lambda^2}{\tau}\left(\frac{dt}{\tau}\right)^{(2/D_F)-1}\Delta c && \text{d}
\end{aligned}\tag{30}$$

Thus, the transport equations (27a-d) are generalized by involving the convective terms $(\hat{V}_F \cdot \nabla)n$, $(\hat{V}_F \cdot \nabla)f$, $(\hat{V}_F \cdot \nabla)m$, $(\hat{V}_F \cdot \nabla)c$ at differentiable scale. Moreover, at the fractal level one specifies new transport mechanisms where the convective effects are balanced by dissipative ones.

Now, the transport equations for the fractal to non-fractal transition are obtained by substracting the relations (29a) and (30a), (29b) and (30b), (29c) and (30c), (29d) and (30d) and using the substitution $\mathbf{V}=\mathbf{V}_D-\mathbf{V}_F$. One gets

$$\begin{aligned}
&\frac{\partial n}{\partial t}+\left(\hat{\mathbf{V}}\cdot\nabla\right)n=\left[D_n-\frac{\lambda^2}{\tau}\left(\frac{dt}{\tau}\right)^{(2/D_F)-1}\right]\Delta n-\chi\nabla\left(n\nabla f\right) && \text{a}\\
&\frac{\partial f}{\partial t}+\left(\hat{\mathbf{V}}\cdot\nabla\right)f=-\frac{\lambda^2}{\tau}\left(\frac{dt}{\tau}\right)^{(2/D_F)-1}\Delta f-\delta mf && \text{b}\\
&\frac{\partial m}{\partial t}+\left(\hat{\mathbf{V}}\cdot\nabla\right)m=\left[D_m-\frac{\lambda^2}{\tau}\left(\frac{dt}{\tau}\right)^{(2/D_F)-1}\right]\Delta m+\mu n-\lambda m && \text{c}\\
&\frac{\partial c}{\partial t}+\left(\hat{\mathbf{V}}\cdot\nabla\right)c=\left[D_c-\frac{\lambda^2}{\tau}\left(\frac{dt}{\tau}\right)^{(2/D_F)-1}\right]\Delta c+\beta f-\gamma n-\alpha c && \text{d}
\end{aligned} \qquad (31)$$

Assuming now both the coherence fractal to non-fractal and harmonic type behavior for the f field the system of equations (31a-d) becomes

$$\begin{aligned}
&\frac{\partial n}{\partial t}=\left[D_n-\frac{\lambda^2}{\tau}\left(\frac{dt}{\tau}\right)^{(2/D_F)-1}\right]\Delta n-\chi\nabla\left(n\nabla f\right) && \text{a}\\
&\frac{\partial f}{\partial t}=-\delta mf && \text{b}\\
&\frac{\partial m}{\partial t}=\left[D_m-\frac{\lambda^2}{\tau}\left(\frac{dt}{\tau}\right)^{(2/D_F)-1}\right]\Delta m+\mu n-\lambda m && \text{c}\\
&\frac{\partial c}{\partial t}=\left[D_c-\frac{\lambda^2}{\tau}\left(\frac{dt}{\tau}\right)^{(2/D_F)-1}\right]\Delta c+\beta f-\gamma n-\alpha c && \text{d}
\end{aligned} \qquad (32)$$

7.3. Non-dimensionalization and parameterization

For us to utilize realistic parameter values, we must first non-dimensionalize the equations in the standard formalism. We therefore rescale the distance with an appropriate length scale L (*e.g.* the maximum invasion distance of the cancer cells at the first stage of invasion, approximately 1 cm), time with τ (*e.g.* the average time taken for mitosis to occur, approximately 8...24 h [130], tumor cell density with n_0, ECM density with f_0, MDE concentration with m_0 and oxygen concentration with c_0 (where n_0, f_0, m_0 and c_0 are appropriate reference variables). Therefore, setting

$$\tilde{n}=\frac{n}{n_0},\quad \tilde{f}=\frac{f}{f_0},\quad \tilde{m}=\frac{m}{m_0},\quad \tilde{c}=\frac{c}{c_0},\quad \tilde{\mathbf{x}}=\frac{\mathbf{x}}{L},\quad \tilde{t}=\frac{t}{t_0}$$

in (27) and dropping the tildes for notational convenience, we obtain the scaled system of equations

$$\begin{aligned}
&\frac{\partial n}{\partial t}=\overbrace{d_n\nabla^2 n}^{\text{random motility}}\ \overbrace{-\rho\nabla\left(n\nabla f\right)}^{\text{haptotaxis}} && \text{a}\\
&\frac{\partial f}{\partial t}=-\overbrace{\eta m f}^{\text{degradation}} && \text{b}\\
&\frac{\partial m}{\partial t}=\overbrace{d_m\nabla^2 m}^{\text{diffusion}}+\overbrace{\kappa n}^{\text{production}}-\overbrace{\sigma m}^{\text{decay}} && \text{c}\\
&\frac{\partial c}{\partial t}=\overbrace{d_c\nabla^2 c}^{\text{diffusion}}+\overbrace{\nu f}^{\text{production}}-\overbrace{\omega n}^{\text{uptake}}-\overbrace{\varphi c}^{\text{decay}} && \text{d}
\end{aligned}\qquad(33)$$

where $d_n=\tau\, D_n-\frac{\lambda^2}{\tau}\left(\frac{dt}{\tau}\right)^{(2/D_F)-1}/L^2$, $\rho=\tau\chi f_0/L^2$, $\eta=\tau m_0\delta$, $d_m=\tau\, D_m-\frac{\lambda^2}{\tau}\left(\frac{dt}{\tau}\right)^{(2/D_F)-1}/L^2$, $\kappa=\tau\mu n_0/m_0$, $\sigma=\tau\lambda$, $d_c=\tau\, D_c-\frac{\lambda^2}{\tau}\left(\frac{dt}{\tau}\right)^{(2/D_F)-1}/L^2$, $\nu=\tau f_0\beta/c_0$, $\omega=\tau n_0\gamma/c_0$, $\phi=\tau\alpha$.

The cell cycle time can be highly variable (particularly the G1 phase) and in fact depends on the specific tumor taken under consideration. As an approximate reference time we take τ = 16 h, halfway between 8...24 h [130]. The cell motility parameter $D_n \sim 10^{-9}$ cm^2 s^{-1} was estimated from available experimental evidence [131]. Tumor cell diameters again will vary depending on the type of tumor being considered but are in the range 10...100 μm [132] with an approximate volume of 10^{-9} to 3×10^{-8} cm^3 [133,134]. We will assume that a tumor cell has the volume 1.5×10^{-8} cm^3 and therefore take $n_0 = 6.7\times10^7$ cells cm^{-3}. The haptotactic parameter $\chi \sim 2600$ cm^2 s^{-1} M^{-1} was estimated to be in line with that calculated in [135] and the parameter $f_0 \sim 10^{-8}$ to 10^{-11} M was taken from the experiments of [136]. We took D_m to be 10^{-9} cm^2 s^{-1}, which is perchance small for a diffusing chemical, but current studies imply that it is in fact a combination of the MDE and MM, and, as a result, the MM degrades and diffuses very little [137]. An *in vivo* estimate for the MDE concentration m_0 is rather difficult to obtain since no value (that we are aware of) has been currently determined and we also know that certain inhibitors (*e.g.* tissue inhibiting metalloproteases) are produced within the ECM which affects the MDE concentration. Plasma levels of specific MDEs have been measured (e.g. MMP-2 [138]) and are approximately 130 ng ml^{-1} with further increases observed in patients with cancer [139]. How is this related to the MDE concentration within the ECM is not clear and we have therefore left this parameter undefined. Estimates for the kinetic parameters μ, λ and δ were not available since these are rather hard to obtain experimentally – and thus we use the values of [135]. The diffusion rate of oxygen through water is $D_c = 10^{-5}$ cm^2 s^{-1} and also, the oxygen consume rate of the cells is 6.25×10^{-17} M cells^{-1} s^{-1} [134]. The background oxygen concentration estimation within the tissue was somehow difficult to be done as it depends on the tissue vascularization. If we set the value of the oxygen concentration in the blood supplying the tumor/tissue to be 0.15 ml O_2 per ml of blood, since we know that 1 M of oxygen occupies 22400 ml then there is 0.15/22400 M O_2 ml^{-1} = 6.7 x 10^{-6} M O_2 ml^{-1}, and since 1 ml = 1 cm^3 then we calculate c_0 = 6.7 x 10^{-6} M O_2 cm^{-3} [140]. Obviously this would be an overestimate, due to the fact that not all of the domain will be fully vascularised but at least we have obtained a reference value. The values of the non-dimensional parameters were given as

$$\begin{array}{ll} d_n = 0.0005, & \text{a} \\ r = 0.01, & \text{b} \\ h = 50, & \text{c} \\ d_m = 0.0005, & \text{d} \\ k = 1, & \text{i} \\ s = 0, & \text{f} \\ d_c = 0.5, & \text{g} \\ n = 0.5, & \text{h} \\ w = 0.57, & \text{i} \\ f = 0.025. & \text{j} \end{array} \tag{34}$$

7.4. Laser beam in a multiscale diffusion cancer-invasion model

7.4.1. Laser as a lorenz system

A laser system is the result of an interaction between the electromagnetic field and the substance, under certain circumstances. The Lorenz form of laser equations may be obtained using a semi-classical reasoning where the environment is analyzed quantically using the formalism of density matrix, and the electromagnetic field is treated classically, by means of Maxwell's equations [141,142]. Here we consider only two energy levels of the involved microscopic systems (atoms, molecules, ions).

The first treatment of a two levels system was made by Bloch who analyzed the interaction of electrons with an oscillatory magnetic field superposed over a static magnetic field, in the framework of a magnetic resonance phenomenon. Due to the similarity of treatments and form of the obtained equations for the laser system, one can say that it forms the Maxwell-Bloch system.

Note that the density matrix method is applied in the treatment of laser systems, no matter how many energy levels, or number of oscillating modes are considered, as well as, in the consequent quantum treatment, where the electromagnetic field is quantized [143,144].

We start by discussing the effect of the electromagnetic field on the atoms of the environment. In the simplest situation, the electric field will induce in each atom an electric dipole whose moment is proportional to the field and is oriented along its direction. Neglecting the vectorial character, we have

$$\mu = \alpha E \tag{35}$$

where α is a constant characteristic to the type of the atom considered. If the concentration of (identical) atoms in the considered environment is N_a, then the polarization vector of the environment, equals the vectorial sum of all the dipole moments from the unit volume, and will be given by

$$P = N_a \mu = \varepsilon_0 \chi E \tag{36}$$

where ε_0 is the empty space permittivity and χ represents the electric susceptibility of the environment.

The problem of the induced dipole moment must be solved quantically using Schrödinger's equation (see paragraph 5.2)

$$i\hbar \frac{\partial \psi}{\partial t} = \hat{H}\psi \tag{37}$$

where $\hat{H}$ is the Hamiltonian operator and ψ the wave function of the atom.

Since a monochromatic field of frequency ω_0, not very intense, interacts with the atom inducing transitions between two of its energetic levels, E_1 and E_2 *i.e.* $E_2 - E_1 = \hbar\omega_0$, it is usual to neglect the other levels and to approximate the atom as a system with two energy levels. If the wave functions of the atom in the two states are ψ_1 and ψ_2, respectively, then we have

$$\psi = C_1\psi_1 + C_2\psi_2 \tag{38}$$

where C_1, C_2 are the time dependent complex amplitude probabilities for the atom to find itself on the energy levels E_1 and E_2, respectively. In other words $\rho_{11}=C_1^*C_1\equiv|C_1|^2$ represents the probability of the atom to find itself in the state ψ_1, and $\rho_{22}=C_2^*C_2\equiv|C_2|^2$ the probability of the atom to find itself in the state ψ_2. The combinations $\rho_{12}=C_1C_2^*$ and $\rho_{21}=C_2C_1^*$ are transition probabilities between the two states. The four numbers ρ_{ij} $(i, j=1, 2)$ forms the so called density matrix for the 2-levels system. The asterisk attached to a parameter means the complex conjugate of the respective parameter. Obviously, $\rho_{21} = \rho^*_{12}$.

The Hamiltonian operator of the system will consist of a sum between the Hamiltonian of the nonperturbed atom $\hat{H}_0$ and a term which describes the interaction of the atom with the field, $\hat{H}'$,

$$\hat{H} = \hat{H}_0 + \hat{H}' \tag{39}$$

The functions ψ_1 and ψ_2 are eigenfunctions of the nonperturbed Hamiltonian, *i.e.*

$$\hat{H}_0\psi_1 = E_1\psi_1, \hat{H}_0\psi_2 = E_2\psi_2, \tag{40}$$

Replacing (38), (39) and (40) into the Schrödinger equation, and after some standard calculus, we get the equations

$$\begin{aligned} i\hbar\dot{C}_1 &= E_1C_1 + H'_{12}C_2, \quad \text{a} \\ i\hbar\dot{C}_2 &= E_2C_2 + H'_{21}C_1, \quad \text{b} \end{aligned} \tag{41}$$

where we introduced the notations

$$\begin{aligned} H'_{12} &= \int \psi_1^*\hat{H}'\psi_2 dV, \text{a} \\ H'_{21} &= \int \psi_2^*\hat{H}'\psi_1 dV, \text{b} \end{aligned} \tag{42}$$

It has been taken into account the orthonormal property of the wave functions ψ_1 and ψ_2

$$\int \psi_i^*\psi_j dV = \delta_{ij} \quad (i,j=1,2) \tag{43}$$

where δ_{ij} *is* Kroeneker's symbol, and the fact that the interaction matrix H'_{ij} has no diagonal elements.

It is common to introduce the following simplification: if one chooses the zero energy value at the center of the interval between the two energies, then they become

$$\begin{aligned} E_2 &= (1/2)\hbar\omega_0, \quad \text{a} \\ E_1 &= -(1/2)\hbar\omega_0, \quad \text{b} \end{aligned} \tag{44}$$

and Eqs. (41) rewrite

$$i\hbar\dot{C}_1 = -\frac{\hbar\omega_0 C_1}{2} + H'_{12}C_2, \quad a$$
$$i\hbar\dot{C}_2 = \frac{\hbar\omega_0 C_2}{2} + H'_{21}C_1, \quad b \tag{45}$$

In quantum mechanics, the electric dipole momentum is calculated as the expectation value of the classical electric dipole momentum $\mu = ex$, where e is the electron charge and x is its displacement along the direction of the electric field.

For an atom in the state ψ this is given by

$$\mu = \int \psi^* ex\psi dV = \left(C_1^* C_2 + C_1 C_2^*\right)\int \psi_1^* ex\psi_2 dV \tag{46}$$

The integral $\mu_{12}=\int\psi_1{}^* ex\psi_2 dV$ represents the electric dipole momentum of the interaction. In the considered approximation, the interaction Hamiltonian is identical to the classical expression of the interaction energy between an electric field and the induced electric dipole: $U = -\mu E$, but with $\mu \to \mu_{12}$ (the dipole momentum of the interaction), i.e.

$$H'_{12} = -\mu_{12}E \tag{47}$$

By choosing a convenient phase relation between the wave functions, we can make μ_{12} real so H'_{12}, H'_{21} to be also real.

Eq. (46) suggests considering the expression

$$X = C_1 C_2^* + C_1^* C_2 \equiv \rho_{12} + \rho_{21} \tag{48}$$

We remark that the polarization (36) is expressed as a function of X through the equation

$$P = N_a \mu_{12} X = N_a \mu_{12} \rho_{12} + c.c. \tag{49}$$

where by c.c. we denote the complex conjugate of the previous expression.

We further consider the combinations

$$Y = i\left(C_1^* C_2 - C_1 C_2^*\right) \equiv -i\left(\rho_{12} - \rho_{21}\right) \tag{50}$$

$$Z = |C_2|^2 - |C_1|^2 \equiv \rho_{22} - \rho_{11} \tag{51}$$

For Z we also have a simple interpretation. If we multiply (51) by N_a we get the expressions $N_a\rho_{22}$ and $N_a\rho_{11}$, which represents the populations from the unit volume of the two levels, in other words, $N_a Z \equiv N$ represents the difference of population between the levels (inversion of population) from the unit volume.

All the three expressions X, Y, Z are functions depending only on time. Their time derivatives are easily calculated using Eqs. (45) and their complex conjugates. The following relations result:

$$\dot{X} = -\omega_0 Y - \frac{i}{\hbar}\left(H'_{12} - H'_{21}\right)Z = -\omega_0 Y \tag{52}$$

$$\dot{Y} = \omega_0 X + \frac{2\mu_{12}}{\hbar} EZ \tag{53}$$

$$\dot{Z} = -\frac{2\mu_{12}}{\hbar} EY \tag{54}$$

where in Eq. (52) the second form was obtained taking into account $H'_{12} = H'_{21}$.

By multiplication of Eq. (52) with $N_a\mu_{12}$, it transforms into an equation for $\dot{P}$, namely

$$\dot{P} = N_a\mu_{12}\dot{X} = N_a\mu_{12}\left(\dot{\rho}_{12} + \dot{\rho}_{21}\right) = N_a\mu_{12}\dot{\rho}_{12} + c.c. \tag{55}$$

The equation for $\dot{\rho}_{12}$ is obtained from Eqs. (52) and (53), taking into account Eqs. (48) and (50). It results

$$\dot{\rho}_{12} = i\omega_0\rho_{12} + i\frac{\mu_{12}}{\hbar} EZ \tag{56}$$

Another equation for polarization is obtained by deriving Eq. (52) once again and using (53). It results

$$\ddot{P} + \omega_0^2 P = -\frac{2\omega_0}{\hbar}\mu_{12}^2 EN \tag{57}$$

It is interesting to note that, in the absence of the electromagnetic field, *i.e.* for $E = 0$, Eq. (57) describes a harmonic oscillator. This is unacceptable, since polarization is induced by the field, so it must attenuate after the field cancels. Physically, it occurs both because of the internal dynamics of the atomic (molecular) systems and of the dephasing between the oscillations of different dipoles by means of their self-interaction or their interaction with the crystal lattice (for a solid environment). This phenomenon is taken into consideration through phenomenological reasonings by introducing an amortization term of the form $\dot{P}/T_2$ in the equation. Eq. (57) transforms into the equation of a forced dumped oscillator:

$$\ddot{P} + \frac{\dot{P}}{T_2} + \omega_0^2 P = -\frac{2\omega_0}{\hbar}\mu_{12}^2 EN \tag{58}$$

Usually, the time T_2 is named *transversal relaxation time*. It is characteristic to the non-diagonal elements of the density matrix, so Eq. (56) must be rewritten

$$\dot{\rho}_{12} + (\gamma_{12} - i\omega_0)\rho_{12} = i\frac{\mu_{12}}{\hbar}EZ \tag{59}$$

where $\gamma_{12} = 1/T_2$.

By multiplication of Eq. (54) with N_a the left side becomes $\dot{N}$. Replacing Y from (52) in (54), we get

$$\frac{\partial N}{\partial t} = \frac{2}{\hbar\omega_0}E\dot{P} \tag{60}$$

Eq. (60) shows that, at the disappearance of the electromagnetic field, the inversion of population must remain constant. However, an electromagnetic field resonant with the considered transition ($\omega \approx \omega_0$) is composed of quanta which can be absorbed by atoms, so may have the effect of a transfer of population between the two levels. It is obvious that, at the canceling of the field, the inversion of population must evolve towards an equilibrium value N^e which is obtained by a process of pumping and by spontaneous relaxation processes. They imply the presence of other energetic levels besides those already considered. We proceed again by phenomenological reasonings. We suppose that this evolution is again exponentially, thus we add a term of the form $(N - N^e)/T_1 \equiv \gamma_{11}(N - N^e)$, where T_1 is named *longitudinal relaxation time* (it is characteristic to the diagonal elements of the density matrix). The equation for the inversion of population becomes

$$\frac{\partial N}{\partial t} + \gamma_{11}(N - N^e) = \frac{2}{\hbar\omega_0}E\dot{P} \tag{61}$$

Eq. (61), together with (55) coupled with (59), or with (58) represents the substance equations. They must be associated with the electromagnetic field equation which we transcribe here

$$\nabla^2\mathbf{E} - \frac{\eta^2}{c^2}\frac{\partial^2\mathbf{E}}{\partial t^2} = \mu_0 \frac{\partial^2\mathbf{P}}{\partial t^2} \tag{62}$$

where $c^2 = 1/\varepsilon_0\mu_0$, $\eta = \sqrt{\varepsilon/\varepsilon_0}$ is the refraction index of the environment (without the contribution of the transition between the two levels), P is the resonant part of the induced polarization and two non-conductive and non-magnetic laser environments were considered. In this case also, the energy losses produced by different mechanisms will be considered phenomenologically by means of an attenuation term introduced in the final form of the equation.

If we have a laser oscillator, the laser medium is placed between two mirrors which form an open cavity, and the oscillations may be triggered only by modes of oscillation characteristic to the cavity. They must satisfy the Helmholtz equation

$$\nabla^2 U + k^2 U = 0 \tag{63}$$

where $k = \eta\omega_c/c$, ω_c being the frequency of the considered mode (index c from cavity). For simplification, in what follows, we consider η=1 (gaseous environment).

We suppose the oscillation is produced on a single mode described by a spatial dependence of the form $W(x, y, z)$. This dependence will characterize both the field and polarization, so we can write

$$\mathbf{E} = \tilde{\mathbf{E}}(t)W(x,y,z)\exp(i\omega_c t) + c.c. \tag{64}$$

and

$$\mathbf{P} = \tilde{\mathbf{P}}(t)W(x,y,z)\exp(i\omega_c t) + c.c. \tag{65}$$

respectively, where $\tilde{\boldsymbol{E}}(t)$ and $\tilde{\boldsymbol{P}}(t)$ are slowly time-varying (complex) amplitudes.

Since the vectors $\boldsymbol{E}$ and $\boldsymbol{P}$ have the same directions, we can neglect the vectorial aspect. Introducing (64) and (66) in (62), applying the slowly varying amplitude approximation, i.e. taking $d^2\tilde{E}/dt^2 \approx 0$, $d^2\tilde{P}/dt^2 \approx -\omega_c P$ and having in view that $W(x, y, z)$ satisfies Eq. (63), we get

$$\frac{d\tilde{E}}{dt} = \frac{i\omega_c}{2\varepsilon_0}\tilde{P} \tag{66}$$

It is necessary to include the loss by radiation which are due to, in the first place, mirrors imperfections. We do that by introducing a term $\kappa\tilde{E}$ in the left hand side, so the field equation becomes

$$\frac{d\tilde{E}}{dt}+\kappa\tilde{E}=\frac{i\omega_c}{2\varepsilon_0}\tilde{P} \tag{67}$$

The equation for polarization is obtained comparing Eqs. (49) and (65). It results

$$\rho_{12}=\frac{\tilde{P}W\exp(i\omega_c t)}{N_a\mu_{12}}, \quad \text{a}$$

$$\dot{\rho}_{12}=\frac{1}{N_a\mu_{12}}\left(\frac{d\tilde{P}}{dt}+i\omega_c\tilde{P}\right)W\exp(i\omega_c t) \quad \text{b} \tag{68}$$

which introduced in Eq. (59), and after multiplying both sides with $W^{*}\exp(-i\omega_c t)$ and integrating over the entire volume (mode) of the cavity, leads to

$$\frac{d\tilde{P}}{dt}+(\gamma_{12}+i(\omega_c-\omega_0))\tilde{P}=\frac{i\mu_{12}^2}{\hbar}\tilde{E}N \tag{69}$$

where, in the left hand side, we introduced the first term from (64) and where N represents the inversion of population from the volume occupied by the cavity mode, defined by the relation

$$N=\frac{N_a\int(\rho_{22}-\rho_{11})W^{*}WdV}{\int W^{*}WdV} \tag{70}$$

We then introduce the relations for the field and polarization (64) and (65) into the equation for the inversion (61). Using the approximation $\dot{P}=i\omega_c P$, neglecting the rapidly varying terms (which contain $\exp(\pm 2i\omega_c t)$), multiplying by WW^{*} and integrating over the volume of the cavity, we get

$$\frac{\partial N}{\partial t}+\gamma_{11}(N-N^{e})=\frac{2iA_0}{\hbar}(\tilde{E}^{*}\tilde{P}-\tilde{E}\tilde{P}^{*}) \tag{71}$$

where N is given by Eq. (70). We consider $\omega_c/\omega_0=1$, and note

$$A_0=\frac{N_a\int(W^{*}W)^2dV}{\int W^{*}WdV} \tag{72}$$

Eqs. (67), (69) and (71) form the Bloch-Maxwell system and describe a unimodal laser oscillator. If we make the change of variables

$$\begin{aligned} t &= \frac{1}{\gamma_{12}} t', && \text{a} \\ \tilde{E} &= \frac{i\hbar\gamma_{12}}{2\sqrt{A_0}\mu_{12}} A, && \text{b} \\ \tilde{P} &= \frac{\hbar\gamma_{12}\varepsilon_0\kappa}{\omega_c\sqrt{A_0}\mu_{12}} R, && \text{c} \\ N - N^e &= \frac{2\hbar\gamma_{12}\varepsilon_0\kappa}{\omega_c\mu_{12}^2} n && \text{d} \end{aligned} \tag{73}$$

the Bloch-Maxwell equations gets the form of the Lorenz system. They become

$$\begin{aligned} \frac{dA}{dt'} &= -\sigma A + \sigma R && \text{a} \\ \frac{dR}{dt'} &= rA - R(1 + i\Omega) - An && \text{b} \\ \frac{dn}{dt'} &= -bn + \frac{1}{2}(AR^* + A^*R) && \text{c} \end{aligned} \tag{74}$$

where the following notations were used

$$\begin{aligned} \sigma &= \frac{\kappa}{\gamma_{12}}, && \text{a} \\ \Omega &= \frac{\omega_c - \omega_0}{\gamma_{12}}, && \text{b} \\ b &= \frac{\gamma_{11}}{\gamma_{12}}, && \text{c} \\ r &= \frac{\omega_c\mu_{12}^2 N^e}{2\varepsilon_0\hbar\kappa\gamma_{12}} && \text{d} \end{aligned} \tag{75}$$

In the form (74) the equations make up a complex Lorenz system. This was discussed in detail in the paper [145]. The complex Lorenz system transforms into the well known real Lorenz system if the resonance is exact ($\omega_c = \omega_0 \Rightarrow \Omega = 0$) and the phases of the amplitudes $\tilde{E}$ and $\tilde{P}$ are chosen so the functions A and R to be real

$$\begin{aligned} \frac{dA}{dt'} &= \sigma(R - A) && \text{a} \\ \frac{dR}{dt'} &= A(r - n) - R && \text{b} \\ \frac{dn}{dt'} &= AR - bn && \text{c} \end{aligned} \tag{76}$$

The fact that a unimodal laser oscillator is described by the Lorenz system was remarked for the first time by Haken (Haken, 1975). Therefore, it was demonstrated that the immense variety of dynamical behaviors, including the chaotic ones, presented by a Lorenz system, must be expected to occur in a laser. Among the first who reported Lorenz type chaotic behaviors in a laser, were Weiss and Brock [147].

Equations of the same form are obtained also when one considers a travelling wave laser, such as laser amplificators where the wave passes only one time the environment, or a circular unidirectional laser [148,149].

7.5. A chaotic multi-scale cancer-invasion model

From the non–dimensional space-time model (33), discretization was performed by neglecting all the spatial derivatives resulting in the following simple 4D temporal dynamical system

$$\begin{aligned} &\frac{dn}{dt}=0 && a\\ &\frac{df}{dt}=-\eta mf && b\\ &\frac{dm}{dt}=\kappa n-\sigma m && c\\ &\frac{dc}{dt}=\nu f-\omega n-\varphi c && d \end{aligned} \tag{77}$$

When simulated, the temporal system (77) with the set of parameters (34) exhibits a virtually linear temporal behavior with almost no coupling between the four concentrations that have very different quantitative values (all phase plots between the four concentrations, not shown here, are virtually one-dimensional). To see if a modified version of the system (77) could lead to a chaotic description of tumor growth, and following the method in [150], four new parameters, a_1, a_2, a_3, and a_4 are introduced. The resulting model is

$$\begin{aligned} &\frac{dn}{dt}=0 && a\\ &\frac{df}{dt}=a_1\eta(m-f) && b\\ &\frac{dm}{dt}=f\left(a_3-c\right)-m+a_2\kappa n && c\\ &\frac{dc}{dt}=\nu fm-a_4\varphi c-\omega n && d \end{aligned} \tag{78}$$

The introduction of the parameters (a_1, a_2, a_3, a_4) was motivated by the fact that tumor cell shape represents a visual manifestation of an underlying balance of forces and chemical reactions [151]. Specifically, the parameters represent the following quantities: a_1 = tumor cell volume (proliferation/non-proliferation fraction), a_2 = glucose level, a_3 = number of tumor cells, a_4 = diffusion from the surface (saturation level).

A tumor is composed of proliferating (P) and quiescent (or non-proliferating) (Q) cells. Tumor cells shift from class P to class Q as the tumor grows in size [152]. Model dependence on the ratio of proliferation to non-proliferation is introduced via the first parameter, a_1. The discretization of Eq. (33a) leads to cell density being modeled as a constant in Eq. (78a). Accordingly, cell density does not play a role in the dynamics. In (78) the cell density is re-introduced into the dynamics via the cell number, a_3. The importance of introducing a_3 also appears in connection with the cyclin-dependent kinase (Cdk) inhibitor p27, the level and activity of which increase in response to cell density. Levels and activity of Cdk inhibitor p27 also increase with differentiation following loss of adhesion to the ECM [153].

The ability to estimate the growth pattern of an individual tumor cell type on the basis of morphological measurements should have general applicability in cellular investigations, cell–growth kinetics, cell transformation and morphogenesis [154].

Cell spreading alone is conducive to proliferation and increases in DNA synthesis, indicating that cell morphology is a critical determinant of cell function, at least in the presence of optimal growth factors and extracellular matrix (ECM) binding [155]. The varying morphology of most cells can stimulate cell proliferation through integrin-mediated signaling indicating that cell shape may govern how individual cells will respond to chemical signals [156].

Parameters (a_1, a_2, a_3, a_4), introduced in connection with cancer cells morphology and dynamics could also influence the very important factor chromatin associated with aggressive tumor phenotype and shorter patient survival time.

For computations, the parameters were set to a_1=0.06, a_2=0.05, a_3=26.5 and a_4=40. Small variation of these chosen values would not affect the qualitative behavior of the new temporal model (78). Simulations of (78), using the same initial conditions and the same non-dimensional parameters as before, show chaotic behavior in the form of Lorenz-like strange attractor in the 3D (f–m–c) subspace of the full 4D (n–f–m–c) phase-space (Figs. 13-Figs. 16).

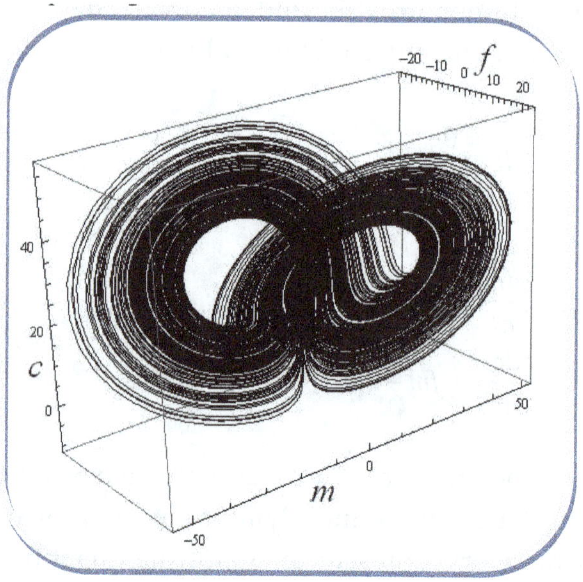

Figure 13. A 3D Lorenz-like chaotic attractor from the modified tumor growth model (78 *b-d*). The attractor effectively couples the MM–concentration f, the MDE–concentration m, and the oxygen concentration c in a mask–like fashion.

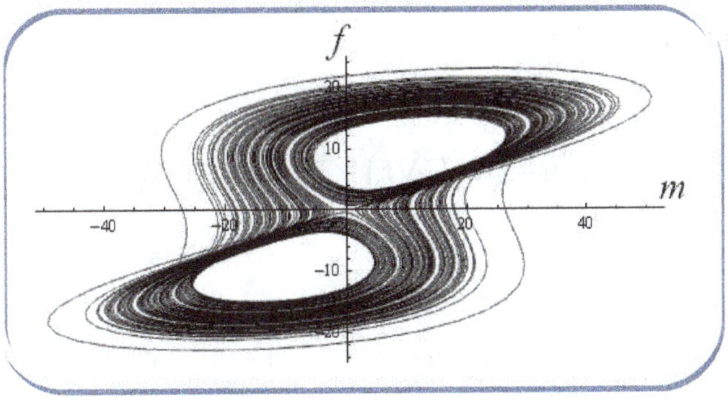

Figure 14. The $m-f$ phase plot of the 3D attractor.

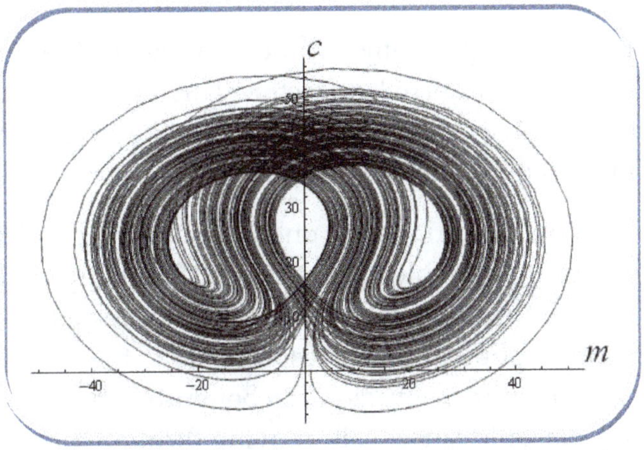

Figure 15. The $m-c$ phase plot of the 3D attractor.

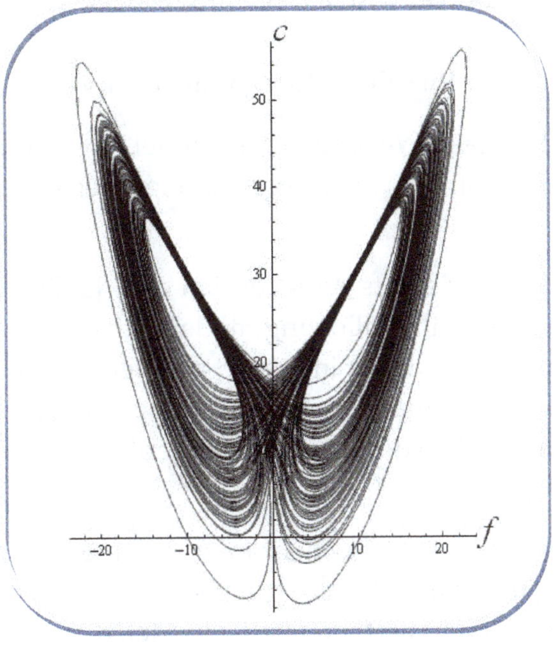

Figure 16. The $f-c$ phase plot of the 3D attractor.

The space-time system of rate PDEs corresponding to the system in (33) provides the following multi-scale cancer invasion model

$$
\begin{aligned}
&\frac{\partial n}{\partial t} = d_n \nabla^2 n - \rho \nabla \cdot (n \nabla f) && \text{a} \\
&\frac{\partial f}{\partial t} = a_1 \eta (m - f) && \text{b} \\
&\frac{\partial m}{\partial t} = d_m \nabla^2 m + a_2 kn + f(a_3 - c) - m && \text{c} \\
&\frac{\partial c}{\partial t} = d_c \nabla^2 c + \nu fm - \omega n - a_4 \varphi c && \text{d}
\end{aligned}
\tag{79}
$$

The new tumor–growth model (79) retains all the qualities of the original model (33) plus includes the temporal chaotic 'butterfly'–attractor. This chaotic behavior may be a more realistic view on the tumor growth, including stochastic–like long–term unpredictability and uncontrollability, as well as sensitive dependence of a tumor growth on its initial conditions.

Now, if we compare (78) with (76) and make a one-to-one correspondence between these two systems of equations, we see that A which is the electric field amplitude corresponds to f, the MM concentration, R which is the polarization amplitude corresponds to m, the MDE concentration and n the inversion of population corresponds to c, the oxygen concentration. Since both systems, the laser and the tumor invasion can be written in the form of a Lorenz system, we can suppose that the metastatic cancer cells moving through the systemic circulation form a coherent wave, i.e. a particular type of chemically pumped (since it may obtain its energy from chemical reactions) laser with oxygen. In the following section we show moreover, that this coherent wave can be identified with a travelling wave laser with oxygen.

7.6. Travelling waves in the multiscale diffusion cancer-invasion model

Let us write the system (28) again. We assume the model refers to the the averaged behavior of the tumor cells in the direction of invasion only and ignores variations in a plane normal to the direction of invasion.

Invasive cells. Since in their experiments Aznavoorian *et al.* [157] reported minimal chemokinetic movement, a key feature of the following model is the absence of the term for random cell motility. Also, we introduce a term of increased proliferation of malignant cells relative to normal cells, $F(n)$, which will be initially modeled as a logistic type growth of the form $k_1 n (k_2 - n)$ which has been shown [158], in order for us to describe adequately the growth of human tumors grown [159].

Extracellular matrix. The motility of extracellular matrix elements, unlike the one of malignant cells and oxygen, is negligible due to the fact that these elements are much longer than cells. The dynamics of connective tissue can therefore be modeled as a simple passive degradation by the activity of the tissue proteases; this proteolysis can now be described by $-G(f, m)$, since it depends on the amount of collagen f still present as well as on the protease m.

Proteases. Proteases generation is narrowly confined to the interface invading tumor and receding connective tissue interface. In some cases it is possible to localize the interstitial collagenase production to the stromal fibroblasts immediately adjacent to the site of tumor invasion, possibly leading to the fact that invasive cells release a stimulus for induction of interstitial collagenase by fibroblasts. Nabeshima *et al.* [160] managed to sequence a tumor cell derived collagenase stimulatory factor. Protease generation located only at the invading front can be explained however in other ways. In their work, Xie *et al.* [161] have revealed that the induction of 92-kd type IV collagenase activity in cultures of A431 human epidermoid carcinoma cells is density dependent. They showed that only dividing cells stained positive when treated with anti-MMP antibodies and as a consequence only noncontact-inhibited tumor cells produce protease. Many proteases are predominantly membrane bound (e.g. uroplasminogen activator), but even when the protease is secreted into the extracellular space, activation occurs only on the cell surface, so as a result the behavior closely resembles that for membrane bound proteases [162]. Therefore protease diffusion in the model is not included. We must then define the function $H(n, f)$ to represent the dependence of this tightly regulated protease production on the local concentrations of the melanoma cells and collagen. In addition we assume that the protease decays linearly, with half-life K.

Oxygen. As in the original model, we presuppose it diffuses into the MM, decays naturally, is consumed by the tumor and for simplicity, oxygen production is proportional to the MM density. Therefore, we introduce the function $I(n, f)$ and c decays linearly, with half-life Λ. The parameter c does not appear anywhere else in the system, so this equation will be easily separated.

Combining all of the above, we are now ready to write the model as

$$\begin{aligned}
&\frac{\partial n}{\partial t} = \overbrace{F(n)}^{\textit{invasive}\ \text{cell proliferation}} - \overbrace{k_3 \frac{\partial}{\partial x}\left(n \frac{\partial f}{\partial x}\right)}^{\textit{haptotactic cell}\ \text{movement}} && \text{a} \\
&\frac{\partial f}{\partial t} = -\overbrace{G(f,m)}^{\textit{proteolysis}} && \text{b} \\
&\frac{\partial m}{\partial t} = \overbrace{H(n,f)}^{\textit{protease production}} - \overbrace{Km}^{\textit{natural decay}} && \text{c} \\
&\frac{\partial c}{\partial t} = \overbrace{d_c \frac{\partial^2 c}{\partial x^2}}^{\textit{diffusion}} + \overbrace{I(f,n)}^{\textit{production and uptake}} - \overbrace{\Lambda c}^{\textit{decay}} && \text{d}
\end{aligned} \tag{80}$$

where F, G, H and I are functions of n, f and m. Compared to previous work on the modeling of cell movement, this model is unusual in that there is no cellular diffusion. This case has been considered previously [163] in the very different context of cellular aggregation, where they obtained conditions for blow-up in the absence of cell kinetics.

Before proceeding further, we must eliminate m from the equations as follows. The time scales associated with protease production and protease decay are much shorter than a typical timescale for the invading cells. Hence writing $H(n, f)=K\overline{H}(n, f)$, where we assume $K>>1$, and multiplying through Eq. (80c) by the small parameter K^{-1}, we deduce that to leading order $m=\overline{H}(n, f)$. Henceforth no reference to m is needed: this expression may be used to eliminate m from Eqs. (80a) and (80*b*). In the same way, writing $I(f, n)=\Lambda\bar{I}(f, n)$, assuming $\Lambda >> 1$ and multiplying through Eq. (80d) by the small parameter Λ^{-1}, we deduce that to leading order $c=\bar{I}(f, n)$. This type of quasi-steady state assumption is a common one in enzyme kinetics [164], and numerical simulations of the four equations (80*a-d*) are in good accordance with the simplified system of two equations; a strong point of the two equation case is that it is amenable to detailed mathematical analysis.

We examine the model using the simple functional forms

$$\begin{aligned} F(n) &= k_1 n(k_2 - n), && \text{a} \\ G(f,m) &= k_4 m f, && \text{b} \\ H(n,f) &= k_5 n f, && \text{c} \\ I(f,n) &= k_6 f - k_7 n && \text{d} \end{aligned} \qquad (81)$$

7.7. Nondimensionalization

After making the substitutions for *F, G, H* and *I* from (81) into equations (80a-d) and eliminating *m* using $m = H(n, f)$ and *c* using $c = I(f, n)$ we nondimensionalize the resulting equations using

$$\tilde{n} = \frac{n}{n^*}, \quad \tilde{f} = \frac{f}{f^*}, \quad \tilde{t} = \frac{t}{T}, \quad \tilde{x} = \frac{x}{L}, \quad L = \left[\frac{k_3}{k_2 k_4 k_5}\right]^{1/2}, \quad T = \frac{1}{k_1 k_2}, \quad n^* = k_2, \quad f^* = \frac{k_1}{k_4 k_5}$$

Dropping tildes for notational convenience then gives rise to the system

$$\begin{aligned} \frac{\partial n}{\partial t} &= n(1-n) - \frac{\partial}{\partial x}\left[n \frac{\partial f}{\partial x}\right] && \text{a} \\ \frac{\partial f}{\partial t} &= -nf^2 && \text{b} \end{aligned} \qquad (82)$$

7.8. Spatially homogeneous system

Setting $\partial/\partial x = 0$ in (82a) we note that the spatially homogeneous system has two steady states:

i. $n = 0$, f arbitrary – this is a continuum of (unstable) steady states parameterized by the (variable) amount of connective tissue in different tissues;

ii. $n = 1$, $f = 0$ – this (stable) steady state corresponds to complete replacement of the normal tissue by invading malignant cells.

With $\partial/\partial x = 0$, (82a) and (82b) can be solved explicitly giving

$$
\begin{aligned}
&n(t) = \left[1 + \exp(-t + c_2)\right]^{-1} && \text{a} \\
&f(t) = [t - c_1 + \log[1 + \exp(-t + c_2)]]^{-1} && \text{b}
\end{aligned}
\qquad (83)
$$

where c_1 and c_2 are arbitrary constants. The behavior as $t \to \infty$ shows that $n \to 1$ and $f \to 0$, hence justifying our classification of the steady state $(n, f) = (1, 0)$ as stable.

7.9. Travelling wave analysis

The invasion process should normally correspond to the travelling wave solutions of the model (82a) and (82b) with the normal tissue steady state $n = 0$ ahead of the wave and the fully malignant state $n = 1, f = 0$ behind the wave. This fact is verified by numerical solutions of (82a) and (82b), which are not detailed here. These travelling wave solutions can be studied analytically using the travelling wave differential equations. We look for constant shape travelling wave front solutions of (82a) and (82b) by setting

$$
\begin{aligned}
&n(x,t) = N(z), && \text{a} \\
&f(x,t) = F(z), && \text{b} \\
&z = x - \xi t && \text{c}
\end{aligned}
\qquad (84)
$$

where ξ is the positive wave speed which has to be determined. When solutions of the type (84) exist, they represent travelling waves moving in the positive x - direction. Substitution of (84) into (82a) and (82b) followed by simple algebraic manipulation gives

$$
\begin{aligned}
&\frac{dN}{dz}\left(-\xi + \frac{2NF^2}{\xi}\right) = N(1-N) - \frac{2N^3F^3}{\xi^2} && \text{a} \\
&\frac{dF}{dz} = \frac{NF^2}{\xi} && \text{b}
\end{aligned}
\qquad (85)
$$

The analysis of (85a,b) involves the study of the (N, F) phase plane. Since we are looking for travelling waves connecting $(1, 0)$ and $(0, \hat{F})$ in the (N, F) phase plane we look for solutions of (85a,b) with boundary conditions

$$
\begin{aligned}
&N(-\infty) = 1, && \text{a} \\
&F(-\infty) = 0, && \text{b} \\
&N(\infty) = 0, && \text{c} \\
&F(\infty) = \hat{F}, && \text{d}
\end{aligned}
\qquad (86)
$$

which requires $(1, 0)$ to have an unstable manifold while $(0, \hat{F})$ must have a stable manifold. In order to study this we look at the stability of the system (85a,b).

7.10. Stability analysis

The steady states (N_0, F_0) of (85a) and (85b) are (0, $\hat{F}$) and (1, 0), where $\hat{F}$ represents a steady states continuum. We study their stability by analyzing the eigenvalues of the stability matrix linearized about the steady states. The eigenvalues about (0, $\hat{F}$) are $-1/\xi$ and 0. The corresponding eigenvectors are (1, $-\hat{F}$) and (0, 1). The negative eigenvalue shows that there is a stable manifold along (1, $-\hat{F}$). The zero eigenvalue represents translations along the steady states continuum.

The eigenvalues about (1, 0) are $1/\xi$ and 0. The eigenvector corresponding to $1/\xi$ is (1, 0) and represents movement along the N axis. The eigenvector corresponding to the zero eigenvalue is (0, 1) which is in the direction normal to the N axis. The trajectory leaving this steady state leaves along the eigenvector corresponding to the zero eigenvalue of the linearized system, as shown by the numerical solutions of (85a,b). This zero eigenvalue is a result of (85b). In order to get a more detailed and clear image of the behavior close to (1, 0) we must look at the nonlinear terms in (85b). One way to do this is to use the techniques of the centre manifold theory which shows that as $z \to -\infty$, $N(z)$ approaches 1 exponentially while $F(z)$ tends zero as z^{-1}.

The existence of an unstable manifold about (0, $\hat{F}$) as $z \to \infty$ and a stable centre manifold about (1, 0) as $z \to -\infty$ is consistent with the existence of a travelling wave orbit connecting the two steady states.

8. Conclusions

1. Cancer cannot be reduced to simple mathematical principles. Its irregular mode of carcinogenesis, erratic tumor growth, variable response to tumoricidal agents, and less-known metastatic patterns constitute highly variable clinical behavior. Characterizing this process requires an accurate understanding of tumor cells and host tissues interactions and ultimately determines prognosis. Applying time-tested and evolving mathematical methods to oncology may provide new methods, with inherent advantages, for the description of tumor behavior, selection of therapeutic modes, prediction of metastatic patterns, and the defining of an inclusive basis for prognostication. Mathematicians describe equations that define tumor growth and behavior, whereas surgeons actively deal with biological processes. Mathematics in oncology applies these principles to clinical settings.

2. The main conclusions of this work are as follows:

   i. mathematics of cancer proves to be chaotic and highly non-linear, justifying the use of space(-time) non-differentiability as a starting base model;

   ii. a chaotic multi-scale cancer-invasion model is manufactured, which embeds a Lorenz attractor in its solutions;

iii. since laser can be expressed as a Lorenz system, we may assume some correspondences between the laser and the above mentioned chaotic multi-scale cancer-invasion model;

iv. the basic model for solid tumor growth admits a travelling wave solution;

v. we suggest that metastatic tumor cells which move through the systemic circulation are similar to a coherent wave, i.e. a travelling wave chemically pumped type laser with oxygen;

vi. we assume the extracellular matrix and in particular, the tumor microenvironment are non-differential media endowed with holographic properties (capacity to memorize, interference abilities and source of forces);

vii. the two well-known phenomena: tumor self-seeding by CTC and hypoxia, in our opinion, both support the idea of complete holography (a hologram which becomes the very object in the particular case of living organisms).

3. Experimentally testable, mathematics applied in oncology may provide a framework to determine clinical outcome on a patient-specific basis and increase the growing awareness that mathematical models help simplify seemingly complex and random tumor behavior.

Author details

Daniel Timofte[1], Lucian Eva[2], Decebal Vasincu[3], Călin Gh. Buzea[4*], Maricel Agop[5,6] and Radu Florin Popa[7]

*Address all correspondence to: calinb2003@yahoo.com

1 Surgery Dept., Hospital "Sf. Spiridon", University of Medicine and Pharmacy "Gr. T. Popa", Iaşi, Romania

2 Emergency Clinical Hospital "Prof. Dr. Nicolae Oblu", Iaşi, Romania

3 Surgery Dept., Biophysical Section, University of Medicine and Pharmacy "Gr. T. Popa", Iaşi, Romania

4 National Institute of Research and Development for Technical Physics – IFT Iasi, Iaşi, Romania

5 Dept. of Physics "Gh. Asachi" Technical University, Iaşi, Romania

6 Lasers, Atoms and Molecules Physics Lab., University of Science and Technology, Lille, France

7 Surgery Dept., University of Medicine and Pharmacy "Gr. T. Popa", Iaşi, Romania

References

[1] Armstrong K., Eisen A., Weber B., *Assessing the Risk of Breast Cancer*. N. Engl. J. Med., 342, 564-571 (2000).

[2] Friedman A., Reitich F., *Analysis of a Mathematical Model for the Growth of Tumors*. J. Math. Biol., 38, 262-284 (1999).

[3] Waliszewski P., Molski M., Konarski J., *On the Holistic Approach in Cellular and Cancer Biology: Nonlinearity, Complexity, and Quasi-determinism of the Dynamic Cellular Network.* J. Surg. Oncol., 68, 70-78 (1998).

[4] Orme M. E., Chaplain M. A. J., *Two-dimensional Models of Tumor Angiogenesis and Antiangiogenesis Strategies.* IMA J. Math. Appl. Med. Biol., 14, 189-205 (1997).

[5] Perumpanani A.J., Sherratt J. A., Norbury J. et al., *Biological Inferences from a Mathematical Model for Malignant Invasion.* Invasion Metastasis, 16, 209-221 (1996).

[6] Secomb et al 2001

[7] Fodde R., Smits R., Clevers H., *Apc, Signal Transduction and Genetic Instability in Colorectal Cancer.* Nature Reviews Cancer, 1, 55 (2001).

[8] Hahn W. C., Counter C. M., Lundberg A. S., Beijersbergen R. L., Brooks M. W., Weinberg R. A., *Creation of Human Tumour Cells with Defined Genetic Elements*. Nature, 400, 464-468 (1999).

[9] Kinzler K., Vogelstein B., *Lessons from Hereditary Colorectal Cancer*. Cell, 87(2), 159-170 (1996).

[10] Olson J., *Bathsheba's Breast: Women and Cancer and History*. John Hopkins University Press, Baltimore, 2002.

[11] Porter R., *The Greatest Benefit to Mankind: A Medical History of Humanity from Antiquity to the Present*. Harper Collins Publishers, London, 1997.

[12] Hajdu S. I., *Greco-roman Thought about Cancer*. Cancer, 100(*10*). 2048-2051 (2004).

[13] Hanahan D., Weinberg A., *The Hallmarks of Cancer*. Cell, 100, 57-70 (2000).

[14] Medema R., Bos J., *The Role of p21ras in Receptor Tyrosine Kinase Signaling*. Crit. Rev. Oncog., 4(*6*) 615-661 (1993).

[15] Weinberg R., *The Retinoblastoma Protein and Cell Cycle Control.* Cell, 81, 323-330 (1995).

[16] Wyllie A., Kerr J., Curri A., *Cell Death: The Significance of Apoptosis.* Int. Rev. Cytol., 68, 251-306 (1980).

[17] Harris C., *P53 Tumor Suppressor Gene: from the Basic Research Laboratory to the Clinic - An Abridged Historical Perspective.* Carcinogenesis, 17, 1187-1198 (1996).

[18] Hayflick L., *Mortality and Immortality at the Cellular Level. A Review.* Biochemistry, 62, 1180-1190 (1997).

[19] Shay J., Bacchetti S., *A Survey of Tolomerase Activity in Human Cancer.* Eur. J. Cancer, 33, 787-791.

[20] Bouck N., Stellmach V., Hsu S., *How Tumors Become Angiogenic.* Adv. Cancer Res. 69, 135–174 (1997).

[21] Hanahan D., Folkman J., *Patterns and Emerging Mechanisms of the Angiogenic Switch During Tumorigenesis.* Cell, 86, 353-364 (1996).

[22] Veikkola T., Alitalo K., *Vegfs and Receptors and Angiogenesis.* Semin. Cancer Biol., 9, 211-220 (1999).

[23] Bull H., Brickell P., Dowd P., *Src-related Protein Tyrosine Kinases are Physically Associated with the Surface Antigen cd36 in Human Dermal Microvascular Endothelial Cells.* FEBS Lett., 351, 41-44 (1994).

[24] Shih T., Lindley C., Bevacizumab, *An Angiogenesis Inhibitor for the Treatment of Solid Malignancies.* Clinical Therapeutics, 28(*11*), 1779-1802 (2006).

[25] Sporn M., *The War on Cancer.* Lancet, 347, 1377-1381 (1996).

[26] Frisch S., Screaton R., *Anoikis Mechanisms.* Current Opinion in Cell Biology, 13(5), 555-562 (2001).

[27] Fidler I., *The Pathogenesis of Cancer Metastasis: The "Seed and Soil" Hypothesis Revisited.* Nature Reviews Cancer, 3 1-6 (2003).

[28] Matrisian L., *The Matrix-degrading Metalloproteinases.* Bioessays, 14, 455-463 (1992).

[29] Mignatti P., Rifkin D., *Biology and Biochemistry of Proteinases in Tumor Invasion.* Physiology Rev., 73, 161-195 (1993).

[30] Araujo R. P., McElwain D. L. S., *A Linear-elastic Model of Anisotropic Tumor Growth.* Eur. J. Appl. Math., 15, 365-384 (2004).

[31] Byrne H. M., Alarcon T., Owen M. R., Webb S. D., Maini P. K., *Modelling Aspects of Cancer Dynamics: A Review.* Phil. Trans. Roy. Soc., A*364*, 1563-1578 (2006).

[32] Adam J., *General Aspects of Modeling Tumor Growth and the Immune Response.* In J. Adam, N. Bellomo (Eds.), *A Survey of Models on Tumor Immune Systems Dynamics*, Birkhäuser, Boston, MA, 1996.

[33] Bellomo N., de Angelis E., Preziosi L., *Multiscale Modelling and Mathematical Problems Related to Tumor Evolution and Medical Therapy.* J. Theor. Med., 5, 111-136 (2003).

[34] Quaranta V., Weaver A. M., Cummings P. T., Anderson A. R. A., *Mathematical Modeling of Cancer: The Future of Prognosis and Treatment.* Clin. Chem. Acta, 357, 173-179 (2005).

[35] Sanga S., Sinek J.P., Frieboes H. B., Fruehauf J. P., Cristini V., *Mathematical Modeling of Cancer Progression and Response to Chemotherapy*. Expert. Rev. Anticancer Ther., 6, 1361-1376 (2006).

[36] Alarcón T., Byrne H. M., Maini P. K., *A Cellular Automaton Model for Tumour Growth in Inhomogeneous Environment.* J. Theor. Biol., 225, 257-274 (2003).

[37] Anderson A. R. A., A *Hybrid Mathematical Model of Solid Tumour Invasion: The Importance of Cell Adhesion.* IMA Math. Appl. Med. Biol., 22, 163-186 (2005).

[38] Mallett D. G., de Pillis L. G., *A Cellular Automata Model of Tumor Immune System Interactions.* J. Theor. Biol., 239, 334-350 (2006).

[39] Abbott R. G., Forrest S., Pienta K. J., *Simulating the Hallmarks of Cancer.* Artif. Life, 12, 617- 634 (2006).

[40] Mansury Y., Kimura M., Lobo J., Deisboeck T.S., *Emerging Patterns in Tumor Systems: Simulating the Dynamics of Multicellular Clusters with an Agent-based Spatial Agglomeration Model.* J. Theor. Biol., 219, 343-370 (2002).

[41] Byrne H. M., Chaplain M. A. J., *Growth of Necrotic Tumors in the Presence and Absence of Inhibitors.* Math. Biosci., 135, 187-216 (1996).

[42] Byrne H. M., Chaplain M. A. J., *Modelling the Role of Cell-cell Adhesion in the Growth and Development of Carcinomas.* Math. Comput. Model., 24, 1-17 (1996).

[43] Greenspan H. P., *On the Growth and Stability of Cell Cultures and Solid Tumors.* J. Theor. Biol., 56, 229-242 (1976).

[44] Araujo R., McElwain D., *A History of the Study of Solid Tumour Growth: The Contribution of Mathematical Modelling.* Bull. Math. Biol., 66, 1039-1091 (2004).

[45] Byrne H. M., Matthews P., *Asymmetric Growth of Models of Avascular Solid Tumors: Exploiting Symmetries.* IMA J. Math. Appl. Med. Biol., 19, 1-29 (2002).

[46] Chaplain M. A. J., Ganesh M., Graham I. G., *Spatio-temporal Pattern Formation on Spherical Surfaces: Numerical Simulation and Application to Solid Tumour Growth.* J. Math. Biol., 42, 387-423 (2001).

[47] Cristini V., Lowengrub J. S., Nie Q., *Nonlinear Simulation of Tumor Growth.* J. Math. Biol., 46, 191-224 (2003).

[48] Li X., Cristini V., Nie Q., Lowengrub J. S., *Nonlinear Three Dimensional Simulation of Solid Tumor Growth. Disc.* Cont. Dyn. Sys., B7, 581-604 (2007).

[49] Ambrosi D., Mollica F., *On the Mechanics of a Growing Tumor.* Int. J. Eng. Sci., 40, 1297-1316 (2002).

[50] Ambrosi D., Preziosi L., *On the Closure of Mass Balance Models for Tumor Growth.* Math. Mod. Meth. Appl. Sci., 12, 737-754 (2002).

[51] Ambrosi D, Guana F., *Stress-modulated Growth.* Math. Mech. Solids (in press) doi: 10.1177/1081286505059739.

[52] Araujo R. P., McElwain D. L. S., *A Mixture Theory for the Genesis of Residual Stresses in Growing Tissues.* (II) *Solutions to the Biphasic Equations for a Multicell Spheroid.* SIAM J. Appl. Math., 66, 447-467 (2005).

[53] Jones A. F., Byrne H. M., Gibson J. S., Dold J. W., A *Mathematical Model of the Stress Induced During Avascular Tumor Growth.* J. Math. Biol., 40, 473-499 (2000).

[54] Roose T., Netti P. A., Munn L. L., Boucher Y., Jain R., *Solid Stress Generated by Spheroid Growth Estimated Using a Linear Poroelastic Model.* Microvasc. Res., 66, 204-212 (2003).

[55] Byrne H., Preziosi L., Modelling *Solid Tumour Growth Using the Theory of Mixtures.* Math. Med. Biol., 20, 341-366 (2003).

[56] Chaplain M. A. J., Graziano L., Preziosi L., *Mathematical Modelling of the Loss of Tissue Compression Responsiveness and its Role in Solid Tumour Development.* Math. Med. Biol., 23, 192-229 (2006).

[57] Zheng X., Wise S. M., Cristini V., *Nonlinear Simulation of Tumor Necrosis, Neo-vascularization and Tissue Invasion Via an Adaptive Finite Element/Level Set Method.* Bull. Math. Biol., 67, 211-259 (2005).

[58] Anderson A. R. A., Chaplain M. A. J., *Continuous and Discrete Mathematical Models of Tumor-induced Angiogenesis.* Bull. Math. Biol., 60, 857-900 (1998).

[59] Sinek J., Frieboes H., Zheng X., Cristini V., *Two-dimensional Chemotherapy Simulati-ons Demonstrate Fundamental Transport and Tumor Response Limitations Involving Nanoparticles.* Biomed. Microdev., 6, 197-309 (2004).

[60] Cristini V., Frieboes H. B., Gatenby R., Caserta S., Ferrari M., Sinek J., *Morphological Instability and Cancer Invasion.* Clin. Cancer Res., 11, 6772-6779 (2005).

[61] Frieboes H. B., Zheng X., Sun C-H., Tromberg B., Gatenby R., Cristini V., *An Integrated Computational/Experimental Model of Tumor Invasion.* Cancer Res., 66, 1597-1604 (2006).

[62] Hogea C. S., Murray B. T., Sethian J. A., *Simulating Complex Tumor Dynamics from Avascular to Vascular Growth Using a General Level-set Method.* J. Math. Biol., 53, 86-134 (2006).

[63] Macklin P., Lowengrub J., *Nonlinear Simulation of the Effect of Microenvironment on Tumor Growth.* J. Theor. Biol., 245, 677-704 (2007).

[64] Wise S. M., Lowengrub J. S., Frieboes H. B., Cristini V., *Three Dimensional Diffuse-interface Simulation of Multispecies Tumor Growth- I: Numerical Method.* Bull. Math. Biol. (in review).

[65] Frieboes H. B., Lowengrub J. S., Wise S., Zheng X., Macklin P., Bearer E. L., Cristini

V., *Computer Simulation of Glioma Growth and Morphology*. NeuroImage, 37, S59-S70 (2007).

[66] Waliszewski P., Molski M., Konarski J., *Self-similarity, Collectivity and Evolution of Fractal Dynamics During Retinoid-induced Differentiation of Cancer Cell Population*. Fractals, 7, 139-149 (1999).

[67] Waliszewski P., Konarski J., Molski M., *On the Modification of Fractal Self-space During Cell Differentiation or Tumor Progression*. Fractals, 8, 195-203 (2000).

[68] Waliszewski P., Molski M., Konarski J., *On the Relationship Between Fractal Geometry of Space and Time in which a Cellular System Exists and Dynamics of Gene Expression*. Acta Biochimol., 48, 209-220 (2001).

[69] Nottale L., *Fractal Space-Time and Microphysics: Towards a Theory of Scale Relativity*. World Scientific, Singapore, 1993.

[70] Guarini G., Onofri E., Menghetti E., *New Horizons in Medicine. The Attractors*. Recenti Prog. Med., 84(9), 618-623 (1993).

[71] Sedivy R., *Fractal Tumours: Their Real and Virtual Images*. Wien Klin. Wochenschr., 108(17), 547-551 (1996).

[72] Sedivy R., Windischberger C., *Fractal Analysis of a Breast Carcinoma – Presentation of a Modern Morphometric Method*. Wien Klin. Wochenschr., 148(14), 335-337 (1998).

[73] Gabor D., *A New Microscopic Principle*. Nature, 161, 777-778 (1948).

[74] Leith E. N., Upatnieks J., *Reconstructed Wavefronts and Communication Theory*. Journal of the Optical Society of America, 52, 1123-30 (1962).

[75] Leith E. N., Upatnieks J., *Wavefront Reconstruction with Continuous-tone Objects*. Journal of the Optical Society of America, 53, 1377-81 (1963).

[76] Leith E.N., Upatnieks J., *Wavefront Reconstruction with Diffused Illumination and Three-dimensional Objects*. Journal of the Optical Society of America, 54, 1295-301 (1964).

[77] Smith H.M. (Ed.), *Holographic Recording Materials*. Springer-Verlag, Berlin, 1977.

[78] Bjelkhagen H. I., *Silver Halide Materials for Holography & Their Processing*. Springer-Verlag, Berlin, 1993.

[79] Pennington K. S., Harper J. S., Laming F. P., *New Phototechnology Suitable for Recording Phase Holograms and Similar Information in Hardened Gelatine*. Applied Physics Letters, 18, 80-84 (1971).

[80] Bartolini R. A., *Photoresists, in: Holographic Recording Materials*. In *Topics in Applied Physics*, Vol. 20, H. M. Smith (Ed.), Springer-Verlag, Berlin, 1977, pp. 209-27.

[81] Booth B. L., *Photopolymer Laser Recording Materials*. Journal of Applied Photographic Engineering, 3, 24-30 (1977).

[82] Smothers W. K., Monroe B. M., Weber A. M., Keys D. E., *Photopolymers for Holography*. In *Practical Holography IV*, Proceedings of the SPIE, Vol. 1212, S. A. Benton (Ed.), SPIE, Bellingham, 1990, pp. 20-29.

[83] Lin L. H., Beauchamp H. L., *Write-read-erase in Situ Optical Memory Using Thermoplastic Holograms.* Applied Optics, 9, 2088–92 (1970).

[84] Urbach J. C., *Thermoplastic Hologram Recording*. In: *Holographic Recording Materials, Topics in Applied Physics*, Vol. 20, H. M. Smith (Ed.), Springer-Verlag, Berlin, 1977, pp. 161–207.

[85] Huignard J. P., Micheron F., *High Sensitivity Read-write Volume Holographic Storage in Bi12SiO20 and Bi12GeO20 Crystals*. Applied Physics Letters, 29, 591–593 (1976).

[86] Huignard J. P., *Phase Conjugation, Real Time Holography and Degenerate Four-wave Mixing in Photoreactive BSO Crystals*. In *Current Trends in Optics*, Arecchi & Aussenegg (Eds.), Taylor & Francis, London, 1981, pp. 150–60.

[87] Svoboda K. K., Fischman D. A., Gordon M. K., *Embryonic Chick Corneal Epithelium: A Model System for Exploring Cell-matrix Interactions*. Developmental Dynamics, 237, 2667-2675 (2008).

[88] Giannoni E., Buricchi F., Grimaldi G. et al., *Redox Regulation of Anoikis: Reactive Oxygen Species as Essential Mediators of Cell Survival.* Cell Death and Differentiation, 15, 867-878 (2008).

[89] Chiarugi P., Giannoni E., *Anoikis: A Necessary Death Program for Anchorage-dependent Cells.* Biochemical Pharmacology, 76, 1352-1364 (2008).

[90] Framson P. E., Sage E. H., *SPARC and Tumor Growth: Where the Seed Meets the Soil.* Journal of Cellular Biochemistry, 92, 679-690 (2004).

[91] Pupa S. M., Menard S., Forti S., Tagliabue E., *New Insights into the Role of Extracellular Matrix During Tumor Onset and Progression.* Journal of Cellular Physiology, 192, 259-267 (2002).

[92] Hofmann U. B., Houben R., Brocker E. B., Becker J. C., *Role of Matrix Metallo-proteinases in Melanoma Cell Invasion*. Biochimie, 87, 307-314 (2005).

[93] von Kempen L. C., Ruiter D. J., van Muijen G. N., Coussens L. M., *The Tumor Microenvironment: a Critical Determinant of Neoplastic Evolution*. European Journal of Cell Biology, 82, 539-548 (2003).

[94] Lopez-Otin C., Palavalli L. H., Samuels Y., *Protective Roles of Matrix Metallo-proteinases: From Mouse Models to Human Cancer.* Cell Cycle, 8, 3657-3662 (2009).

[95] Wu Y., Zhou B. P., *Inflammation: A Driving Force Speeds Cancer Metastasis.* Cell Cycle, 8, 3267-3273 (2009).

[96] Nottale L., *Scale Relativity and Fractal Spece-Time,* Imperial College Press, London (2011).

[97] Mandelbrot B., *The Fractal Geometry of Nature*, W.H. Freeman, New York (1983).

[98] Casian Botez I., Agop M., Nica P., Paun V., Munceleanu G.V., *Conductive and Convective Type Behavior at Nano-Time Scales*, J. Comput. Theor. Nanosci., 7, 2271-228 (2010).

[99] Agop M., Forna N., Casian-Botez I., Bejenariu I. C., *New Theoretical Approach of the Physical Processes in Nanostructures*, J. Comput. Theor. Nanosci., 5, 483-489 (2008).

[100] Munceleanu G.V., Paun V.P., Casian Botez I., Agop M., *The Microscopic-macroscopic Scale Transitions through a Chaos Scenario in the Fractal Space-Time Theory*, International Journal of Bifurcation and Chaos, 21, 603-618 (2011).

[101] Luis G., *Complex Fluids*, Springer, Vol. 415 (1993).

[102] Mitchell M., *Complexity: A Guided Tour*, Oxford Univ. Press, Oxford (2009).

[103] Thomas Y. Hou, Multiscale Phenomena in Complex Fluids: Modeling, Analysis and Numerical Simulations, World Scientific Publishing Company, Singapore (2009).

[104] Landau L., Lifshitz E.M., *Fluid Mechanics*, 2-nd Edition, Butterworth Heinemann, Oxford (1987).

[105] Birlescu V.S., Agop M., M. Craus, *Computational Properties of a Fractal Medium*, International Journal of Quantum Information, 12, 4 (2014).

[106] Norton L., Massagué J., *Is Cancer a Sisease of Self-seeding*? Nature Medicine, 12, 875-878 (2006).

[107] Kim Mi-Y., Oskarsson T., Acharyya S., Nguyen D. X., Zhang X. H.-F., Norton L., Massagué J., *Tumor Self-seeding by Circulating Cancer Cells*. Cell, 139, 1315-1326 (2009).

[108] Enderling H., Anderson A.R.A., Chaplain M.A.J., Beheshti A., Hlatky L., Hahnfeldt P., *Paradoxical Dependencies of Tumor Dormancy and Progression on Basic Cell Kinetics*. Cancer Research, 69(22), 8814-8821 (2009).

[109] Vaupel P., Kelleher D. K., Hockel M., *Oxygen Status of Malignant Tumors: Pathogenesis of Hypoxia and Significance for Tumor Therapy*. Semin. Oncol., 28, 29-35 (2001).

[110] Vaupel P., *The Role of Hypoxia-induced Factors in Tumor Progression*. Oncologist, 9, Suppl. 5, 10-17 (2004).

[111] Harris A. L., *Hypoxia – A Key Regulatory Factor in Tumour Growth*. Nat. Rev. Cancer, 2, 38-47 (2002).

[112] Kimbro K.S., Simons J.W., *Hypoxia-inducible Factor-1 in Human Breast and Prostate Cancer*. Endocr. Relat. Cancer, 13, 739-749 (2006).

[113] O'Donnell J. L., Joyce M. R., Shannon A. M., Harmey J., Geraghty J., Bouchier-Hayes D., *Oncological Implications of Hypoxia Inducible Factor-1alpha (HIF-1alpha) Expression*. Cancer Treat. Rev., 32, 407-416 (2006).

[114] Lovestam M.–A, Agardh E., *Photocoagulation of Diabetic Macular Oedema: Complications and Visual outcome*. Acta Ophthalmolm. Scand., 78, 667-671 (2000).

[115] McCarty C. A., McKay R., Keeffe J. E., *Management of Diabetic Retinopathy by Australian Ophthalmologists*. Working Group on Evaluation of the NHMRC Retinopathy Guideline Distribution, National Health and Medical Research Council. Clin. Exp. Ophthalmol., Vol. 28, 2000, pp. 107-112.

[116] Fong D. S., Ferris F. L. III, Davis M. D., Chew E. Y., *Causes of Severe Visual Loss in the Early Treatment Diabetic Retinopathy Study*. ETDRS Report no. 24, *Early Treatment Diabetic Retinopathy Study Research Group*. Am. J. Ophthalmol., 127, 137-141 (1999).

[117] Dogru M., Nakamura M., Inoue M., Yamamoto M., *Long-term Visual Outcome in Proliferative Diabetic Retinopathy Patients after Panretinal Photo-coagulation*. Jpn. J. Ophthalmol., 43, 217-224 (1999).

[118] Yu D.-Y., Cringle S. J., Su E., Yu P. K., Humayun M. S., Dorin G., *Laser-Induced Changes in Intraretinal Oxygen Distribution in Pigmented Rabbits*. Invest. Ophthalmol. Vis. Sci., 46(3), 988-999 (2005).

[119] Cringle S..J., Yu D.-Y., Yu P. K., Su E.-N., *Intraretinal Oxygen Consumption in the Rat in Vivo*. Invest. Ophthalmol. Vis. Sci., 43, 1922-1927(2002).

[120] Cringle S..J., Yu D-Y., A *Multi-layer Model of Retinal Oxygen Supply and Consumption Helps Explain the Muted Rise in Inner Retinal PO_2 During Systemic Hyperoxia*. Comp. Biochem Physiol., 132, 61-66 (2002).

[121] Carter S.B., *Principles of Cell Motility: The Direction of Cell Movement and Cancer Invasion*. Nature, 208, 1183-187 (1965).

[122] Quigley J. P., Lacovara J., Cramer E. B., *The Directed Migration of B-16 Melanoma-cells in Response to a Haptotactic Chemotactic Gradient of Fibronectin*. J. Cell Biol., 97, A450-A451 (1983).

[123] Lacovara J., Cramer E. B., Quigley J. P., *Fibronectin Enhancement of Directed Migration of B16 Melanoma Cells*. Cancer Res., 44, 1657-1663 (1984).

[124] McCarthy J. B., Furcht L.T., *Laminin and Fibronectin Promote the Directed Migration of B16 Melanoma Cells in Vitro*. J. Cell Biol., 98, 1474-1480 (1984).

[125] Klominek J., Robert K. H., Sundqvist K.-G., *Chemotaxis and Haptotaxis of Human Malignant Mesothelioma Cells: Effects of Fibronectin, Laminin, Type IV Collagen, and an Autocrine Motility Factor-like Substance*. Cancer Res., 53, 4376-4382 (1993).

[126] Lawrence J. A., Steeg P. S., *Mechanisms of Tumour Invasion and Metastasis*. World J. Urol., 14, 124-130 (1996).

[127] Debruyne P.R., Bruyneel E.A., Karaguni I.-M. et al., *Bile Acids Stimulate Invasion and Haptotaxis in Human Corectal Cancer Cells through Activation of Multiple Oncogenic Signalling Pathways*. Oncogene, 21, 6740-6750 (2002).

[128] Stetler-Stevenson W. G., Hewitt R., Corcoran M., *Matrix Metallo-proteinases and Tumour Invasion: From Correlation to Causality to the Clinic*. Cancer Biol., 7, 147-154 (1996).

[129] Chambers A. F., Matrisian L. M., *Changing Views of the Role of Matrix Metalloproteinases in Metastasis.* J. Natl. Cancer Inst., 89, 1260-1270 (1997).

[130] Calabresi P., Schein P. S. (Eds.), *Medical Oncology.* 2nd Edn., McGraw-Hill, New York, 1993.

[131] Bray D., *Cell Movements.* Garland Publishing, New York, 1992.

[132] Melicow M. M., *The Three-steps to Cancer: A New Concept of Carcinogenesis.* J. Theor. Biol., 94, 471-511 (1982).

[133] Folkman J., Hochberg M., *Self-regulation of Growth in Three Dimensions.* J. Exp. Med., 138, 745-753 (1973).

[134] Casciari J. J., Sotirchos S. V., Sutherland R. M., *Variation in Tumour Cell Growth Rates and Metabolism with Oxygen-concentration, Glucose-concentration and Extracellular pH.* J. Cell. Physiol., 151, 386-394 (1992).

[135] Anderson A.R.A., Chaplain M.A.J., Newman E.L., Steele R.J.C., Thompson A.M., *Mathematical Modelling of Tumour Invasion and Metastasis.* J. Theor. Med., 2, 129-154 (2000).

[136] Terranova V.P., Diflorio R., Lyall R.M., Hic S., Friesel R., Maciag T., *Human Endothelial Cells are Chemotactic to Endothelial Cell Growth Factor and Heparin.* J. Cell Biol., 101, 2330-2334 (1985).

[137] Hotary K., Allen E.D., Punturieri A., Yana I., Weiss S.J., *Regulation of Cell Invasion and Morphogenesis in a 3-dimensional Type I Collagen Matrix by Membrane-type Metalloproteinases 1, 2 and 3.* J. Cell Biol., 149, 1309-1323 (2000).

[138] Zervoudaki A., Economou E., Pitsavos C., Vasiliadou K. et al., *The Effect of Ca2+ Channel Antagonists on Plasma Concentrations of Matrix Metallopro ases in Metastasis.* J. Natl. Cancer Inst., 89, 1260-1270 (1997).

[139] Johansson N., Ahonen M., Kahari V.-M., Matrix Metalloproteinases in Tumour Invasion. Cell. Mol. Life Sci., 57, 5-15 (2000).

[140] Sherwood L., *Human Physiology: From Cells to Systems.* 4th Ed., Pacific Grove, California, Brooks/Cole, 2001.

[141] Loudon R., *The Quantum Theory of Light.* 2-nd Ed., Oxford Univ. Press, Oxford, 1983.

[142] Sargent M. III, Scully M. O., Lamb W. E. Jr., *Laser Physics.* Addison Wesley Publ. Co., Reading, Massachusetts, 1977.

[143] Pantel R. H., Puthoff H. E., *Fundamentals of Quantum Electronics.* John Wiley & Sons, Inc., NewYork, 1969.

[144] Nussenzveig H. M., *Introduction to Quantum Optics.* Gordon and Breach Sci. Publ., London, 1973.

[145] Fowler A. C., Gibbon J. D., Mcguinness M. J., Physica, D4, 139 (1982).

[146] Haken H., Phys. Lett., A *53*, 77 (1975).

[147] Weiss B. O., Brock J., Phys. Rev. Lett., 57, 2804 (1986).

[148] Milonni P. W., Shih M.-L., Ackerhalt J. R., *Chaos in Laser-Matter Interactions*. World Sci. Publ. Co., Singapore, 1987.

[149] Newell A. C., Moloney J. W., *Nonlinear Optics*. Addison Wesley Publ. Co., Reading, Massachusetts, 1992.

[150] Ivancevic T. T., Bottema M. J., Jain L. C., *A Theoretical Model of Chaotic Attractor in Tumor Growth and Metastasis*. arXiv:0807.4272v1 [q-bio.CB] 27 Jul 2008.

[151] Olive L. C., Durand R. E., *Drug and Radiation Resistance in Spheroids: Cell Contact and Kinetics*. Cancer Metastasis Rev., 13, 121 (1994).

[152] Kozusko F., Bourdeau M., *A Unified Model of Sigmoid Tumour Growth Based on Cell Proliferation and Quiescence*. Cell Prolif., 40(6), 824-834 (2007).

[153] Chu I. M., Hengst L., Slingerland J. M., *The Cdk Inhibitor p27 in Human Cancer: Prognostic Potential and Relevance to Anticancer Therapy*. Nature Rev. Cancer, 8, 253-287 (2008).

[154] Castro M.A.A., Klamt F., Grieneisen V.A., Grivicich I., Moreira J.C.F., *Gompertzian Growth Pattern Correlated with Phenotypic Organization of Colon Carcinoma, Malignant Glioma and Non-small Cell Lung Carcinoma Cell Lines*. Cell Prolif. 36(2), 65-73 (2003).

[155] Ingber D. E., *Fibronectin Controls Capillary Endothelial Cell Growth by Modulating Cell Shape*. Proc. Natl. Acad. Sci., USA, 87 (1990).

[156] Boudreau N., Jones P. L., *Extracellular Matrix and Integrin Signaling: The Shape of Things to Come*. Biochem. J., 339 (1999).

[157] Aznavoorian M., Stracke L., Krutzsch H., Schiffman E., Liotta L. A., *Signal Transduction for Chemotaxis and Haptotaxis by Matrix Molecules in Tumour Cells*. J. Cell Biol., 110, 1427-1438 (1990).

[158] Vaidya V.G., Alexandro F.J. Jr., *Evaluation of Some Mathematical Models for Tumour Growth*. Int. J. Biomed. Comput., 13, 19-35 (1982).

[159] Schwartz M., *A Biomathematical Approach to Clinical Tumour Growth*. Cancer, 14, (1961).

[160] Nabeshima K., Lane W.S., Biswas C., *Partial Sequencing and Characterisation of the Tumor Cell-derived Collagenase Stimulatory Factor*. Arch. Biochem. Biophys., 285, 90-96 (1991).

[161] Xie B., Bucana C. D., Fidler I. J., *Density-dependent Induction of 92-kd Type Type-IV Col-*

lagenase Activity in Cultures of A431 Human Epidermoid Carcinoma Cells. Am. J. Pathol., 144, 1958-1967 (1994).

[162] Werb Z., *ECM and Cell Surface Proteolysis: Regulating Cellular Ecology*. Cell, 91, 439-442 (1997).

[163] Rascle M., Ziti C., *Finite Time Blow-up in Some Models of Chemotaxis.* J. Math. Biol., 33, 388-414 (1995).

[164] Murray J. D., *Mathematical Biology*. Springer, Berlin, 1990.

[165] Hariharan P, *Basics of Holography,* Cambridge University Press, New York, 2002.

4

Physical Vacuum is a Special Superfluid Medium

V.I. Sbitnev

1. Introduction

A dramatic situation in physical understanding of the nature emerged in the late of 19th century. Observed phenomena on micro scales came into contradiction with the general positions of classical physics. It was a time of the origination of new physical ideas explaining these phenomena. Actually, in a very short period, postulates of the new science, quantum mechanics were formulated. The Copenhagen interpretation was first who proposed an ontological basis of quantum mechanics [1]. These positions can be stated in the following points: (a) the description of nature is essentially probabilistic; (b) a quantum system is completely described by a wave function; (c) the system manifests wave–particle duality; (d) it is not possible to measure all variables of the system at the same time; (e) each measurement of the quantum system entails the collapse of the wave function.

Can one imagine a passage of a quantum particle (the heavy fullerene molecule [2], for example) through all slits, in once, at the interference experiment? Following the Copenhagen interpretation, the particle does not exists until it is registered. Instead, the wave function represents it existence within an experimental scene [3].

Another interpretation was proposed by Louis de Broglie [4], which permits to explain such an experiment. In de Broglie's wave mechanics and the double solution theory there are two waves. There is the wave function that is a mathematical construct. It does not physically exist and is used to determine the probabilistic results of experiments. There is also a physical wave guiding the particle from its creation to detection. As the particle moves from a source to a detector, the particle perturbs the wave field and gets a reverse effect from it. As a result, the physical wave guides the particle along some optimal trajectory up to its detection.

A question arises, what is the de Broglie physical wave? Recently, Couder and Fort [5] has executed the experiment with the classical oil droplets bouncing on the oil surface. A remarkable observation is that an ensemble of the droplets passing through the barrier having two gates shows the interference fringes typical for the two slit experiment. Their explanation is that the droplet while moving on the surface induces on this surface the weak Faraday waves. The latter provide the guidance conditions for the droplets. In this perspective, we can draw conclusion that the de Broglie physical wave can be represented by perturbations of the ether when the particle moves through it. In order to describe behavior of such an unusual medium we shall use the Navier-Stokes equation with slightly modified some terms. As the final result we shall get the Schrödinger equation.

In physical science of the New time the assumption for the existence of the ether medium was originally used to explain propagation of light and the long-range interactions. As for the propagation of light, the wave ideas of Huygens and Fresnel require the existence of a continuous intermediate environment between a source and a receiver of the light - the light-bearing ether. It is instructive to compare here the two opposite doctrines about the nature of light belonging to Sir Isaac Newton and Christian Huygens. Newton maintained the theory that the light was made up of tiny particles, corpuscles. They spread through an empty space in accordance with the law of the classical mechanics. Christian Huygens (a contemporary of Newton), believed that the light was made up of waves vibrating up and down perpendicularly to the direction of its propagation, as waves on a water surface. One can imagine all space populated everywhere densely by Huygens's vibrators. All vibrators are silent until a wave reaches them. As soon as a wave front reaches them, the vibrators begin to radiate waves on the frequency of the incident wave. So, the infinitesimal volume δV is populated by infinite amount of the vibrators with frequencies of visible light. These vibrators populate the ether facilitating propagation of the light waves through the space.

In order to come to idea about existence of the intermediate medium (ether) that penetrates overall material world, we begin from the fundamental laws of classical physics. Three Newton's laws first published in Mathematical Principles of Natural Philosophy in 1687 [6] we recognize as basic laws of physics. Namely: (a) the first law postulates existence of inertial reference frames: an object that is at rest will stay at rest unless an external force acts upon it; an object that is in motion will not change its velocity unless an external force acts upon it. The inertia is a property of the bodies to resist to changing their velocity; (b) the second law states: the net force applied to a body with a mass M is equal to the rate of change of its linear momentum in an inertial reference frame

$$\vec{F} = M\vec{a} = M\frac{d\vec{v}}{dt}; \tag{1}$$

(c) the third law states: for every action there is an equal and opposite reaction.

Leonard Euler had generalized the second Newton's laws on motion of deformable bodies [7]. We rewrite this law for such media. Let the deformable body be in volume ΔV and has the mass M. We divide Eq. (1) by ΔV and determine the time-dependent mass density $\rho_M = M\Delta V$ [8]. In this case, we shall understand $\vec{F}$ as force per volume. Then the second law takes a form:

$$\vec{F} = \frac{d\rho_M \vec{v}}{dt} = \rho_M \frac{d\vec{v}}{dt} + \vec{v}\frac{d\rho_M}{dt}. \quad (2)$$

The total derivatives in the right side can be written through partial derivatives:

$$\frac{d\rho_M}{dt} = \frac{\partial \rho_M}{\partial t} + (\vec{v}\nabla)\rho_M. \quad (3)$$

$$\frac{d\vec{v}}{dt} = \frac{\partial \vec{v}}{\partial t} + (\vec{v}\nabla)\vec{v}. \quad (4)$$

Eq. (3) equated to zero is seen to be the continuity equation. As for Eq. (4) we may rewrite the rightmost term in detail

$$(\vec{v}\nabla)\vec{v} = \nabla\frac{v^2}{2} - [\vec{v}\times[\nabla\times\vec{v}]]. \quad (5)$$

As follows from this formula, the first term, multiplied by the mass, is gradient of the kinetic energy. It represents a force applied to the fluid element for its shifting on the unit of length, δS. The second term gives acceleration of the fluid element directed perpendicularly to the velocity $\vec{v}$. Let the fluid element move along some curve in 3D space. Tangent to the curve in each point points to orientation of the body motion. In turn, the vector $\vec{\omega}=[\nabla\times\vec{v}]$ is perpendicular to the plane, within which an arbitrarily small segment of the curve is situated. This vector characterizes a quantitative measure of the vortex motion. It is called *vorticity*. The vector product $[\vec{v}\times\vec{\omega}]$ is perpendicular to the both vectors $\vec{v}$ and $\vec{\omega}$. It shows the acceleration of the fluid element

The term (5) entering in the Navier-Stokes equation [9, 10] is responsible for emergence of vortex structures. The Navier-Stokes equation stems from Eq. (2) if we omit the rightmost term, representing the continuity equation, and specify forces in this equation in detail:

$$\rho_M\left(\frac{\partial \vec{v}}{\partial t} + (\vec{v}\nabla)\vec{v}\right) = \frac{\vec{F}}{\Delta V} + \mu(t)\nabla^2\vec{v} - \rho_M\nabla\left(\frac{P}{\rho_M}\right) \quad (6)$$

This equation contains two modifications represented in the two last terms from the right side: the dynamic viscosity μ depends on time and the rightmost term has a slightly modified view, namely $\rho_M\nabla(P/\rho_M)=\nabla P - P\nabla\ln(\rho_M)$. This modification will be important for us when we shall begin to derive the Schrödinger equation. However first we shall examine the Helmholtz vortices with time dependent dynamical viscosity.

2. Vortex dynamics

The second term from the right in Eq. (6) represents the viscosity of the fluid (μ is the dynamic viscosity, its units are N·s/m^2=kg/(m·s)). Let us suppose that the fluid is ideal, barotropic, and the mass forces are conservative [10]. At assuming that the external force is conservative, we apply to this equation the operator curl. We get right away the equation for the vorticity:

$$\frac{\partial \vec{\omega}}{\partial t} + (\vec{\omega} \cdot \nabla)\vec{v} = \nu(t)\nabla^2 \vec{\omega}. \quad (7)$$

Here $\nu(t)=\mu(t)/\rho_M$ is the kinematic viscosity. Its dimensionality is m^2/s what corresponds to dimensionality of the diffusion coefficient. The rightmost term describes dissipation of the energy stored in the vortex. As a result, the vortex with the lapse of time will disappear.

With omitted the term from the right (i.e., $\nu=0$) the Helmholtz theorem reads: (i) if fluid particles form, in any moment of the time, a vortex line, then the same particles support the vortex line both in the past and in the future; (ii) ensemble of the vortex lines traced through a closed contour forms a vortex tube. Intensity of the vortex tube is constant along its length and does not change in time. The vortex tube (a) either goes to infinity by both endings; (b) or these endings lean on walls of bath containing the fluid; (c) or these endings are locked to each on other forming a vortex ring.

Assuming that the fluid is a physical vacuum, which meets the requirements specified earlier, we must say that the viscosity vanishes. In that case, the vorticity $\vec{\omega}$ is concentrated in the center of the vortex, i.e., in the point. Mathematical representation of the vorticity is δ- function. Such singularity can be a source of possible divergences of computations in further.

We shall not remove the viscosity. Instead of that, we hypothesize that even if there is an arbitrary small viscosity, because of the zero-point oscillations in the vacuum, the vortex does not disappear completely. The vortex can be a long-lived object. The foundation for that hypothesis is the observation (performed by French scientific team [5, 11, 12]) of behavior of the droplets moving on the oil surface, on which the waves Faraday exist. Here an important moment is that the Faraday waves are supported slightly below the super-critical threshold. Due to this trick the droplets can live on the oil surface arbitrary long, before they disappear in the oil. The Faraday waves that are supported near the super-critical threshold may play a role analogous to the zero-point oscillations of the vacuum.

Observe that the bouncing droplet simulates some aspects of quantum mechanics, stimulating theoretical investigations in this area [13-20]. It is interesting to note in this place that Grössing considers a quantum particle as a dissipative phase-locked steady state, where an amount of zero-point energy of the wave-like environment is absorbed by the particle, and then during a characteristic relaxation time is dissipated into the environment again [14].

Here we shall give a simple model of such a picture. Let us look on the vortex tube in its cross-section which is oriented along the axis z and its center is placed in the coordinate origin of the plane (x, y). Eq. (7), written down in the cross-section of the vortex, is as follows

$$\frac{\partial \omega}{\partial t} = \nu g(t)\left(\frac{\partial^2 \omega}{\partial r^2} + \frac{1}{r}\frac{\partial \omega}{\partial r}\right). \tag{8}$$

We do not write a sign of vector on top of ω since ω is oriented strictly along the axis z. We introduce time-dependent the kinematic viscosity. For the sake of simplicity, let it be looked as

$$g(t) = \cos(\Omega t + \phi) = \frac{e^{\mathrm{i}(\Omega t+\phi)} + e^{-\mathrm{i}(\Omega t+\phi)}}{2} \tag{9}$$

where Ω is an oscillation frequency and ϕ is the uncertain phase.

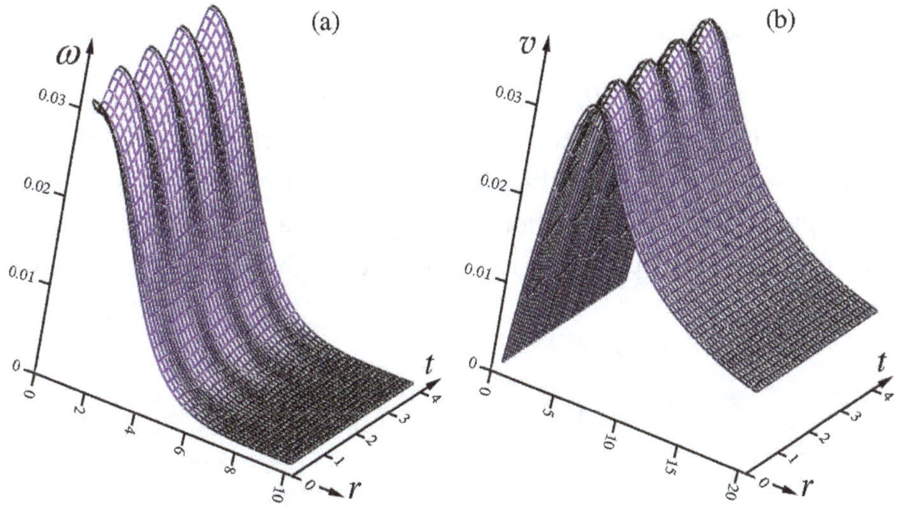

Figure 1. Vorticity $\omega(r,\ t)$ and velocity $v(r,t)$ as functions of r and t for $\Gamma=1$, $\nu=1$, $\Omega=\pi$, and n=16. These parameters are conditional in order to show clearly oscillations of the vortex in time. The solution does not decay with time.

Solution of the equation (8) in this case is as follows

$$\omega(r,t) = \frac{\Gamma}{4\pi(\nu/\Omega)(\sin(\Omega t+\phi)+n)}\exp\left\{-\frac{r^2}{4(\nu/\Omega)(\sin(\Omega t+\phi)+n)}\right\}. \tag{10}$$

Here Γ is the integration constant having dimension m^2/s. An extra number $n > 1$. It prevents appearance of singularity in the cases when $\sin(\Omega t + \phi)$ tends to-1. This function at choosing the parameters $\Gamma=1$, $\nu=1$, $\Omega=\pi$, and n=16 is shown in Fig. 1(a).

The velocity of the fluid matter around the vortex results from the integration of the vorticity function

$$v(r,t) = \frac{1}{r}\int_0^r \omega(r',t)r'dr' = \frac{\Gamma}{2\pi r}\left(1-\exp\left\{-\frac{r^2}{4(\nu/\Omega)(\sin(\Omega t+\phi)+n)}\right\}\right). \tag{11}$$

Fig. 1(b) shows behavior of this function at the same input parameters.

In particular, for n=0 and $\Omega\, t<<1$ this solution is close to the Lamb–Oseen vortex solution [21]

$$\omega(r,t) = \frac{\Gamma}{4\pi\nu t}e^{-r^2/4\nu t}, \tag{12}$$

$$v(r,t) = \frac{\Gamma}{2\pi r}\left(1-e^{-r^2/4\nu t}\right). \tag{13}$$

As seen from here, the Lamb–Oseen solution decays with time since the viscosity ν>0.

One can see from the solutions (10) and (11), depending on the distance to the center the functions $\omega(r,\, t)$ and $v(r,t)$ show typical behavior for the vortices. The both functions do not decay with time, however. Instead of that, they demonstrate pulsations on the frequency Ω. Amplitude of the pulsations is the smaller, the larger value of the parameter n. At n tending to infinity the amplitude of the pulsations tends to zero. At the same time the vortex disappears entirely.

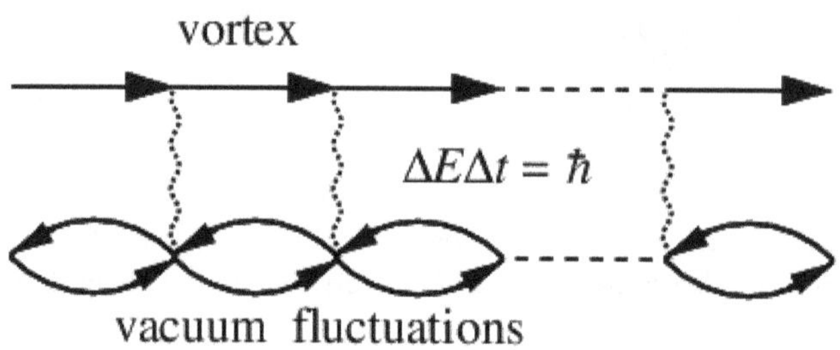

Figure 2. Periodic energy exchange between the vortex and vacuum fluctuations

The undamped solution was obtained thanks to assumption, that the kinematic viscosity is a periodic function of time, namely, $\nu\, g(t)$=$\nu\cos(\Omega\, t+\phi)$. The viscosity in the quantum realm is not a good concept, however. Most likely, it manifests itself through interaction of the quantum object with vacuum fluctuations. According to Eq. (9), there are half-periods when the energy of the vortex is lost at scattering on the vacuum fluctuations, and there are other half-periods when the vacuum returns this energy to the vortex, Fig. 2. On the whole, the viscosity of the fluid medium, within which the vortex tube evolves, in the average remains at zero. It can mean that this medium is superfluid. Such a scenario is not unusual. For example, at transition of helium to the superfluid phase [22] coherent Cooper pairs of electrons arise through the exchange by phonons. This attraction is due to the electron–phonon interaction. The phonons are thermal excitations of a lattice. In that case, they play a role of the background medium.

Qualitative view of the vortex tube in its cross-section is shown in Fig. 3. Values of the velocity v are shown by grey color ranging from light grey (minimal velocities) to dark grey (maximal ones). A visual image of this picture can be a hurricane (tropical cyclone [23]) shown from the top. In the center of the vortex, a so-called eye of the hurricane (the vortex core) is well viewed. Here it looks as a small light grey disk, where the velocities have small values. In the very center of the disk, in particular, the velocity vanishes. Observe that in the region of the hurricane eye a wind is really very weak, especially near the center. This is in stark contrast to conditions in the region of the eyewall, where the strongest winds exist (in Fig. 3 it looks as a dark grey annular region enclosing the light grey inner area). The eyewall of the vortex tube (a zone where the velocity reaches maximal values) has the nonzero radius.

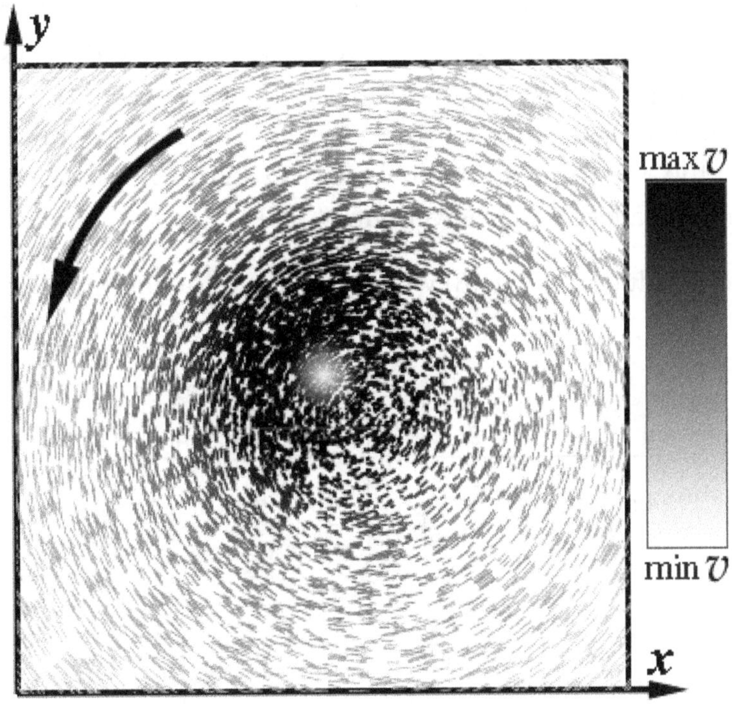

Figure 3. Cross-section of the vortex tube in the plane (x, y). Values of the velocity v are shown in grey ranging from light grey (min v) to dark grey (max v). Density of the pixels represents magnitude of the vorticity ω. Core of the vortex is well visible in the center.

Let us find the radius of the vortex core. In order to evaluate this radius we equate to zero the first derivative by r of equation (11)

$$\frac{\exp\left\{-\frac{r^2}{4(\nu/\Omega)(\sin(\Omega t+\phi)+n)}\right\}\left(2\frac{r^2}{4(\nu/\Omega)(\sin(\Omega t+\phi)+n)}+1\right)-1}{2\pi r^2}=0. \tag{14}$$

The radius is a root of this equation

$$r_{\mathrm{v}} = 2\sqrt{a_0(n + \sin(\Omega t + \phi))} \cdot \sqrt{\frac{\nu}{\Omega}}. \tag{15}$$

Here $a_0 \approx 1.2564312$ is a root of the equation $\ln(2a_0 + 1) - a_0 = 0$. One can see, that r_v is an oscillating function. The larger Ω, the more quickly the vortex trembles. As Ω increases, the vortex radius decreases. However it grows with increasing the number n. Let us evaluate the radius r_v at choosing the viscosity ν equal to $\bar{\nu} = \hbar / 2m$. Here $\bar{\nu}$ is the diffusion coefficient of the Brownian sub-quantum particles wandering in the Nelson's aether [24], see Appendix A. In the case of electron $\bar{\nu} = \hbar / 2m \approx 5.79 \bullet 10^{-5}$ m²/s. As for Ω, let it be equal to $2mc^2/\hbar$, or approximately $1.6 \bullet 10^{21}$ radians per second for electron. Here c is the speed of light and m is the electron mass. Then we have $(\nu/\Omega)^{1/2} \approx 1.93 \bullet 10^{-13}$ m. This length is seen to be smaller then the Compton wavelength, $\lambda_C = 2.426 \bullet 10^{-12}$ m, in about 12 times. So, for choosing $n \approx 31$ we find from Eq. (15) that the radius of the vortex is about the Compton wavelength. From the above one can see that, on a distance about the Compton wavelength, virtual particles can be involved into a vorticity dancing around the electron core, by polarizing the electron charge. This dancing happens at trembling motion of the electron with the frequency $\Omega = 2mc^2/\hbar$. That oscillating motion has a deep relation to the so-called "Zitterbewengung" [25].

One can give a general solution of Eq. (8) which has the following presentation

$$\omega(r,t) = \frac{\Gamma}{4\pi\left(\int_0^t \nu(\tau)d\tau + \sigma^2\right)} \exp\left\{-\frac{r^2}{4\left(\int_0^t \nu(\tau)d\tau + \sigma^2\right)}\right\}, \tag{16}$$

$$v(r,t) = \frac{\Gamma}{2\pi r}\left(1 - \exp\left\{-\frac{r^2}{4\left(\int_0^t \nu(\tau)d\tau + \sigma^2\right)}\right\}\right). \tag{17}$$

The viscosity function $\nu(t)$ is a quasi-periodic function or even is represented by a color noise. The integral of the viscosity function memorizes integrally character of the viscosity of the medium. Due to this memory effect, the vortex may live a long enough. As for interpretation of these solutions with the quantum-mechanical point of view, we may say that there exists a regular exchange by quanta with the vacuum fluctuations, Fig. 2. The integral accumulates all cases of the exchange with the vacuum. The constant σ having dimension of length, prevents appearance of singularities. One can see that even at $\nu \equiv 0$, but $\sigma > 0$, abundance of long-lived vortices can exist in the vacuum. Such vortices are "ghosts" in the superfluid being invisible without interaction.

2.1. Vortex rings and vortex balls

If we roll up the vortex tube in a ring and glue together its opposite ends we obtain a vortex ring. A result of such an operation put into the (x, y) plane is shown in Fig 4. Position of points on the helicoidal vortex ring in the Cartesian coordinate system is given by

$$\begin{cases} x = (r_1 + r_0 \cos(\omega_2 t + \phi_2))\cos(\omega_1 t + \phi_1), \\ y = (r_1 + r_0 \cos(\omega_2 t + \phi_2))\sin(\omega_1 t + \phi_1), \\ z = r_0 \sin(\omega_2 t + \phi_2). \end{cases} \tag{18}$$

Here r_0 is the radius of the tube. And r_1 represents the distance from the center of the tube (pointed in the figure by arrow c) to the center of the torus located in the origin of coordinates (x,y,z). A body of the tube, for the sake of visualization is colored in cyan. Eq. (18) parametrized by t gives a helicoidal vortex ring shown in this figure. Parameters ω_1 and ω_2 are frequencies of rotation along the arrow a about the center of the torus (about the axis z) and rotation along the arrow b about the center of the tube (about the axis pointed by arrow c), respectively. Phases ϕ_1 and ϕ_2 have uncertain quantities ranging from 0 to 2π. By choosing the phases within this interval with a small increment, we may fill the torus by the helicoidal vortices everywhere densely. The vorticity is maximal along the center of the tube. Whereas the velocity of rotation about this center in the vicinity of it is minimal. However, the velocity grows as a distance from the center increases. After reaching of some maximal value the velocity further begins to decrease.

Let the radius r_1 in Eq. (18) tends to zero. The helicoidal vortex ring in this case will transform into a vortex ring enveloping a spherical ball. The vortex ring for the case r_0=4, $r_1 \approx 0$, ω_2=$3\omega_1$, and ϕ_2=ϕ_1=0 drawn by thick curve colored in deep green is shown in Fig. 5(a). Motion of an elementary vortex clot along the vortex ring (along the thick curve colored in deep green) takes place with a velocity

$$\vec{v}_R = \begin{pmatrix} v_{R,x} = -r_0\omega_2 \sin(\omega_2 t + \phi_2)\cos(\omega_1 t + \phi_1) - r_0\omega_1 \cos(\omega_2 t + \phi_2)\sin(\omega_1 t + \phi_1) \\ \quad - r_1\omega_1 \sin(\omega_1 t + \phi_1) \\ v_{R,y} = -r_0\omega_2 \sin(\omega_2 t + \phi_2)\sin(\omega_1 t + \phi_1) + r_0\omega_1 \cos(\omega_2 t + \phi_2)\cos(\omega_1 t + \phi_1) \\ \quad + r_1\omega_1 \sin(\omega_1 t + \phi_1) \\ v_{R,x} = \ r_0\omega_2 \cos(\omega_2 t + \phi_2) \end{pmatrix} \tag{19}$$

The velocity of the clot at the initial time is $v_{R,x}$=0, $v_{R,y}$=$r_0\omega_1$, $v_{R,z}$=$r_0\omega_2$=$3r_0\omega_1$ (the initial point (x, y, z)=(4, 0, 0) is on the top of the ball). We designate this velocity as $\vec{v}_+$. Through the time t=$\pi\omega_1$ the elementary vortex clot returns to the top position. The velocity in this case is equal

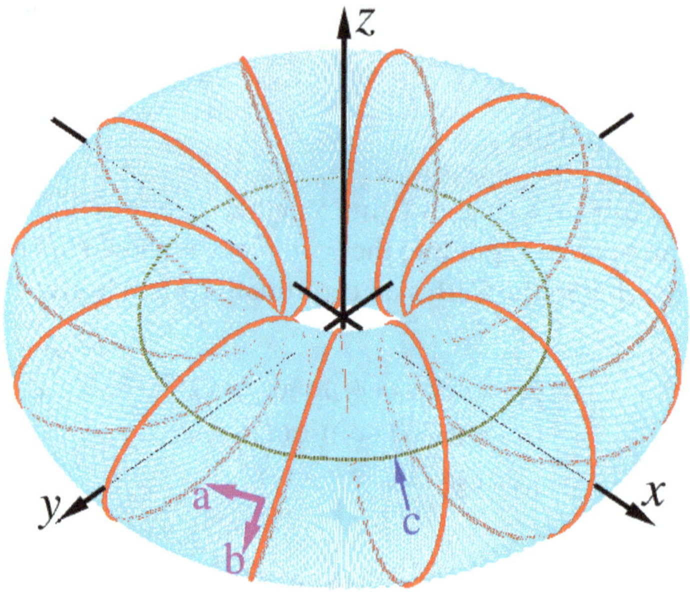

Figure 4. Helicoidal vortex ring: r_0=2, r_1=3, $\omega_2=12\omega_1$, $\phi_2=\phi_1=0$.

to $v_{R,x}=0$, $v_{R,y}=r_0\omega_1$, $v_{R,z}=-r_0\omega_2=-3r_0\omega_1$. We designate this velocity as $\vec{v}_-$. Sum of the two opposite velocities, $\vec{v}_+$ and $\vec{v}_-$, gives the velocity $\vec{v}_0=(0,\ 2r_0\omega_1,\ 0)$. During $t=(1+3k)\pi/3\omega_1$ and $t=(2+3k)\pi/3\omega_1$ (k=1, 2, ⋯) the clot travels through the positions 1 and 2 both in the forward and in backward directions, respectively. In the vicinity of these points the velocities $\vec{v}_+$ and $\vec{v}_-$ yield the resulting velocity $\vec{v}_0$ directed along the circle lying in the plane (x, y).

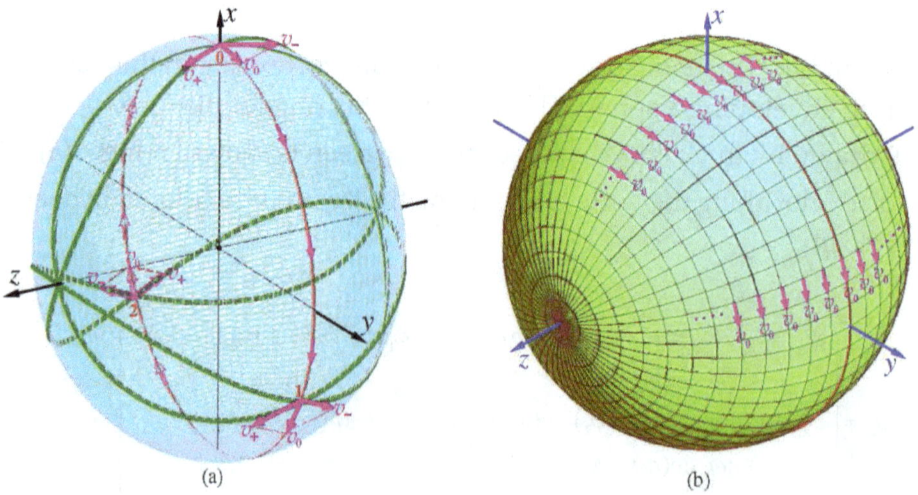

Figure 5. (a) Helicoidal vortex ring (colored in deep green) convoluted into the vortex ball. The input parameters of the ball are as follows r_0=4, r_1=0.01 $\ll$ 1, $\omega_2=3\omega_1$, $\phi_2=\phi_1=0$. The radius r_0 represents a mean radius of the ball, where the velocity v_0 reaches a maximal value. The ratio $\omega_2/\omega_1=3$ was chosen with the aim not to overload the picture by superfluous curves. (b) The vortex ball rotating about axis z with the maximal velocity v_0 that is reached on the surface of the ball.

The ball can be filled everywhere densely by other rings at adding them with other phases ϕ_1 and ϕ_2 ranging from 0 to 2π. The velocity $\vec{v}_0$ for any ring will lie on the same circles centered on the axis z. We see a dense ball that rolls along the axis y, Fig. 5(b). Observe that the ball pulsates on the frequency Ω as it rolls along its path, as it follows from the above computations. Perfect modes describing the rolling ball are spherical harmonics [26].

3.3. Derivation of the Schrödinger equation

The third term in the right side of Eq. (6) deals with the pressure gradient. One can see, however, it is slightly differ from the pressure gradient presented in the customary Navier-Stokes equation [9, 10]. One can rewrite this term in detail

$$\rho_M \nabla\left(\frac{P}{\rho_M}\right) = \nabla P - P\nabla \ln(\rho_M). \tag{20}$$

The first term, ∇P, is the customary pressure gradient represented in the Navier-Stokes equation. Whereas, the second term, $P\nabla \ln(\rho_M)$, is an extra term describing changing the logarithm of the density along increment of length (the entropy increment) multiplied by P. It may mean that change of the pressure is induced by change of the entropy per length, or else by change of the information flow [27, 28] per length. This term has signs typical of the osmotic pressure, mentioned by Nelson [24].

Let us consider in this respect the pressure P in more detail. We shall represent the pressure consisting of two parts, P_1 and P_2. We begin from the Fick's law [14]. The law says that the diffusion flux, J, is proportional to the negative value of the density gradient $J = -D\nabla\rho_M$. Here D is the diffusion coefficient $\bar{v} = \hbar/2m$ [24], see Nelson's definition in Appendix A. Since the term $\bar{v}\nabla J$ has dimension of the pressure, we define P_1 as the pressure having diffusion nature

$$P_1 = \bar{v}\,\nabla\vec{J} = -\frac{\hbar^2}{4m^2}\nabla^2\rho_M. \tag{21}$$

Observe that the kinetic energy of the diffusion flux is $(m/2)(J/\rho_M)^2$. It means that there exists one more pressure as the average momentum transfer per unit area per unit time:

$$P_2 = \frac{\rho_M}{2}\left(\frac{\vec{J}}{\rho_M}\right)^2 = \frac{\hbar^2}{8m^2}\frac{(\nabla\rho_M)^2}{\rho_M}. \tag{22}$$

Now we can see that sum of the two pressures, P_1+P_2, divided by ρ_M gives a term

$$Q_M = \frac{\hbar^2}{8m^2}\left(\frac{\nabla\rho_M}{\rho_M}\right)^2 - \frac{\hbar^2}{4m^2}\frac{\nabla^2\rho_M}{\rho_M}. \tag{23}$$

One can see that accurate to the divisor m this term represents the quantum potential.

To bring the expression (23) to a form of the quantum potential, we need to introduce instead of the mass density ρ_M the probability density ρ according to the following presentation

$$\rho_M = \frac{M}{\Delta V} = \frac{mN}{\Delta V} = m\rho. \tag{24}$$

Here the mass M is a product of an elementary mass m by the number of these masses, N, filling the volume ΔV. Then the mass density ρ_M is defined as a product of the elementary mass m by the density of quasi-particles $\rho = N / \Delta V$. We can imagine the quasi-particle as a long-lived local heterogeneity, which moves with the current velocity $\vec{v}$ and probably has the vorticity. Let us divide the Navier-Stokes equation Eq. (6) by the probability density ρ. We obtain

$$m\left(\frac{\partial\vec{v}}{\partial t} + (\vec{v}\nabla)\vec{v}\right) = \frac{\vec{F}}{N} + \nu(t)\nabla^2 m\,\vec{v} - \nabla Q \tag{25}$$

Here $\vec{F}/N$ is the force per one the quasi-particle. The kinetic viscosity $\nu(t) = \mu(t)/\rho_M$ is represented through the diffusion coefficient $\bar{v} = \hbar/2m$ [24], $\nu(t) = 2\bar{v}\,g(t) = \nu\,g(t)$, $\nu = \hbar/m$, where $g(t)$ is the dimensionless time dependent function. The function Q here is the real quantum potential

$$Q = \frac{\hbar^2}{8m}\left(\frac{\nabla\rho}{\rho}\right)^2 - \frac{\hbar^2}{4m}\frac{\nabla^2\rho}{\rho}. \tag{26}$$

Grössing noticed that the term ∇Q, the gradient of the quantum potential, describes a completely thermalized fluctuating force field [13, 14]. Here the fluctuating force is expressed via the gradient of the pressure divided by the density distribution of sub-quantum particles chaotically moving in the environment. Perhaps, they are virtual particle-antiparticle pairs.

Since the pressure provides a basis of the quantum potential, as was shown above, it would be interesting to interpret an osmotic nature of the pressure [24]. The interpretation can be the following (see Appendix A): a semipermeable membrane where the osmotic pressure manifests itself is an instant, which divides the past and the future (that is, the 3D brane of our being represents the semipermeable membrane in the 4D world). In other words, the thermalized fluctuating force field described by Grössing [13, 14] is asymmetric with respect to the time arrow.

3.1. Transition to the Schrödinger equation

The current velocity v contains two component – irrotational and solenoidal [10] that relate to vortex-free and vortex motions of the medium, respectively. The basis for the latter is the Kelvin-Stokes theorem. Scalar and vector fields underlie of manifestation of the irrotational and solenoidal velocities

$$\vec{v} = \vec{v}_S + \vec{v}_R = \frac{1}{m}\nabla S + \vec{v}_R. \quad (27)$$

Here subscripts S and R hint to scalar and vector (rotational) potentials underlying emergence of these two components of the velocity. These velocities are submitted by the following equations

$$\begin{cases} \left(\nabla \cdot \vec{v}_S\right) \neq 0, & \left[\nabla \times \vec{v}_S\right] = 0, \\ \left(\nabla \cdot \vec{v}_R\right) = 0, & \left[\nabla \times \vec{v}_R\right] = \vec{\omega}. \end{cases} \quad (28)$$

The scalar field is represented by the scalar function S – action in classical mechanics. Both velocities are perpendicular to each other. We may define the momentum and the kinetic energy

$$\begin{cases} \vec{p} = m\vec{v} = \nabla S + m\vec{v}_R, \\ m\dfrac{v^2}{2} = \dfrac{1}{2m}(\nabla S)^2 + m\dfrac{v_R^2}{2}. \end{cases} \quad (29)$$

Now we may rewrite the Navier-Stokes equation (25) in the more detailed form

$$\frac{\partial}{\partial t}\left(\nabla S + m\vec{v}_R\right) + \underbrace{\frac{1}{2m}\nabla\left((\nabla S)^2 + m^2 v_R^2\right) + \left[\vec{\omega} \times \left(\nabla S + m\vec{v}_R\right)\right]}_{(a)} = -\nabla U - \nabla Q + \underbrace{\nu(t)\nabla^2\left(\nabla S + m\vec{v}_R\right)}_{(b)}. \quad (30)$$

Note that the term embraced by the curly bracket (a) stems from $(\vec{v}\,\nabla)\vec{v} = \nabla v^2/2 + [\vec{\omega} \times \vec{v}]$, see Eq. (5). Here we take into account that the external force is conservative, i.e., $\vec{F}/N = -\nabla U$, where U is the potential energy relating to the single quasi-particle. The term embraced by the curly bracket (b) describes the viscosity of the medium. As was said above the viscosity coefficient in the average is equal to zero.

Let us rewrite Eq. (30) by regrouping the terms

$$\nabla\left(\frac{\partial}{\partial t}S+\frac{1}{2m}(\nabla S)^2+\frac{m}{2}\vec{v}_R^2+U+Q-\nu(t)\nabla^2 S\right)=-m\frac{\partial}{\partial t}\vec{v}_R-\left[\vec{\omega}\times\left(\nabla S+m\vec{v}_R\right)\right]+\nu(t)m\nabla^2\vec{v}_R. \quad (31)$$

We assume that fluctuations of the viscosity about zero occur much more frequent, than characteristic time of displacements of the quasi-particles. For that reason, we omit the term $\nu(t)\,\nabla^2 S$ by supposing in the first approximation, that the medium is absolutely superfluid-there are no energy sources and sinks. By multiplying this equation from the left by $\vec{v}_S$ we find that the right part of this equation vanishes since $(\vec{v}_S \cdot \vec{v}_R)=(\vec{v}_S \cdot \vec{\omega})=0$. The left part vanishes if the expression under the brackets is constant. As a result, we come to the following modified Hamilton-Jacobi equation

$$\frac{\partial}{\partial t}S+\frac{1}{2m}(\nabla S)^2+\frac{m}{2}v_R^2+U(\vec{r})-\frac{\hbar^2}{2m}\left(\frac{\nabla^2\rho}{2\rho}-\left(\frac{\nabla\rho}{2\rho}\right)^2\right)=C. \quad (32)$$

The modification is due to adding the quantum potential (26). In this equation, C is an integration constant. We see that the third term in this equation represents energy of the vortex. On the other hand, we can see that the vortex given by Eq. (7) is replenished by the kinetic energy coming from the scalar field S, namely via the term $(\vec{\omega}\cdot\nabla)\vec{v}$. Solutions of these two equations, Eq. (7) and Eq. (32), describing dynamics of the vortex and scalar fields, depend on each other.

Both the continuity equation

$$\frac{\partial\rho}{\partial t}+(\vec{v}\cdot\nabla)\rho=0, \quad (33)$$

which stems from Eq. (3), and the quantum Hamilton-Jacobi equation (32) can be extracted from the following Schrödinger equation

$$\mathrm{i}\hbar\frac{\partial\Psi}{\partial t}=\frac{1}{2m}(-\mathrm{i}\hbar\nabla+m\vec{v}_R)^2\Psi+U(\vec{r})\Psi-C\Psi. \quad (34)$$

The kinetic momentum operator $(-\mathrm{i}\hbar\nabla+m\vec{v}_R)$ contains the term $m\vec{v}_R$ describing a contribution of the vortex motion. This term is analogous to the vector potential multiplied by the ratio of the charge to the light speed, which appears in quantum electrodynamics [29]. Appearance of this term in this equation is conditioned by the Helmholtz theorem.

By substituting into Eq. (34) the wave function Ψ represented in a polar form

$$\Psi=\sqrt{\rho}\cdot\exp\{\,\mathrm{i}S/\hbar\,\} \quad (35)$$

and separating on real and imaginary parts we come to Eqs. (32) and (33). So, the Navier-Stokes equation (6) with the slightly expanded the pressure gradient term can be reduced to the Schrödinger equation if we take into consideration also the continuity equation.

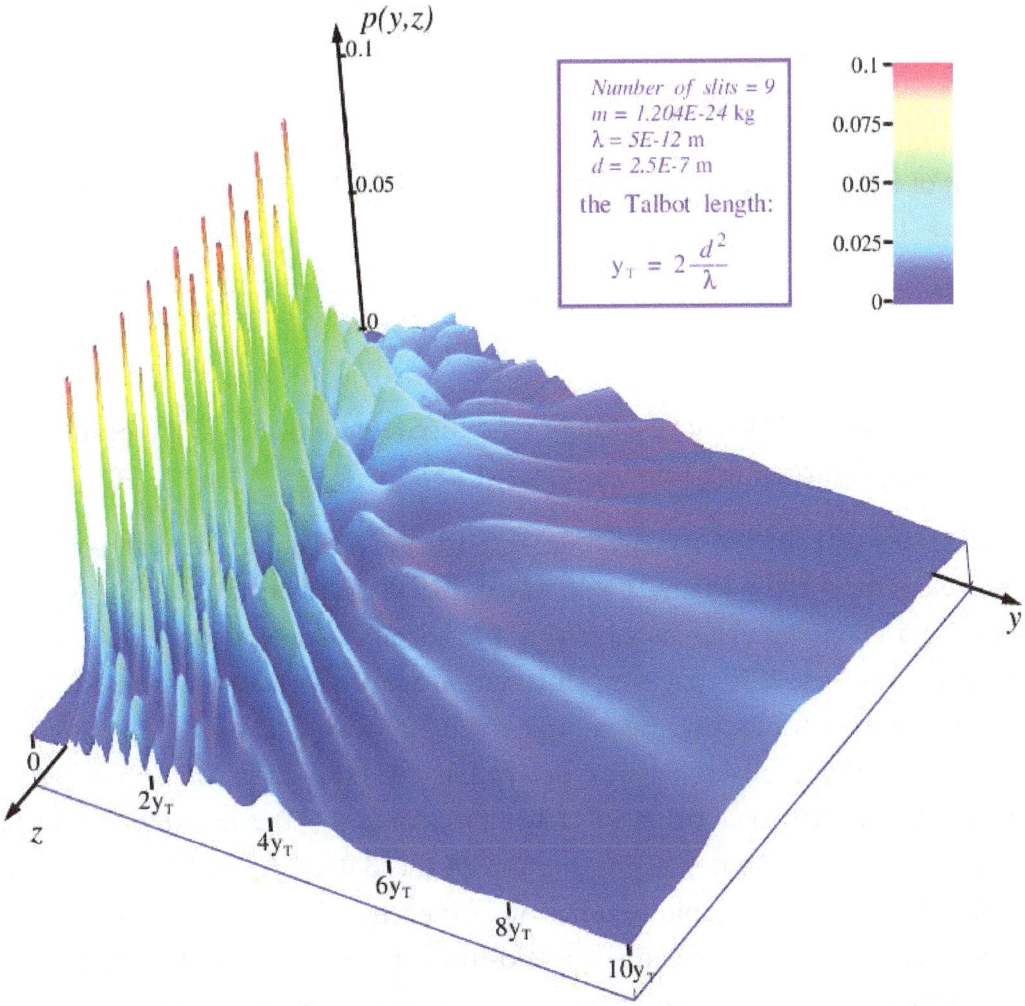

Figure 6. Probability density distribution from scattering the fullerene molecules on the grating containing 9 slits: de Broglie wavelength is 5 pm and the distance between slits is 250 nm.

There are confirmations that the Schrödinger equation is deduced from the Feynman path integral [30, 31]. Therefore, for searching solutions of the Schrödinger equation we may apply the path integral. The solution of the Schrödinger equation (34) with the potential that simulates a grating with N slits has the following view [32]

$$|\Psi(y,z)\rangle = \frac{1}{N\sqrt{1+\mathrm{i}\dfrac{\lambda y}{2\pi b^2}}}\sum_{n=0}^{N-1}\exp\left\{-\frac{\left(z-\left(n-\dfrac{N-1}{2}\right)d\right)^2}{2b^2\left(1+\mathrm{i}\dfrac{\lambda y}{2\pi b^2}\right)}\right\}. \tag{36}$$

Here λ is the de Broglie wavelength, d is the distance between slits, b is the slit width, and **i**=√-1. In this calculation we have used $\lambda = 5\,\text{pm}$, $b = 5 \cdot 10^3 \lambda$, and $d = 5 \cdot 10^4 \lambda = 10b = 250$ nm. By choosing N=9 slits, for example, we find the interference pattern shown in Fig. 6 as the density distribution function [32]. This function is a scalar product of the wave function $|\Psi(y,z)\rangle$, namely:

$$p(y,z) = \langle \Psi(y,z) | \Psi(y,z) \rangle. \tag{37}$$

A useful unit of length at observation of the interference patterns is the Talbot length:

$$y_{\mathrm{T}} = 2\frac{d^2}{\lambda}. \tag{38}$$

This length bears name of Henry Fox Talbot who discovered in 1836 [33] a beautiful interference pattern, named further the Talbot carpet [34, 35].

The particles, incident on the slit grating, come from a distant coherent source. The de Broglie wavelength of the particle, $\lambda = h / p$ (h is the Planck constant, and p is the particle momentum) is a main characteristics binding the corpuscular Newtonian physics with the wave Huygens' physics. It is that we call now the wave-particle dualism. The de Broglie pilot wave being represented by the complex-valued wave function $|\Psi\rangle$ fills all ambient space, except of opaque objects, which determine the boundary conditions. The particle passes from the source to a place of detection along the optimal trajectory, Bohmian trajectory [27]. The equation describing motion of the Bohmian particle can be found, for example, in [36]. There is the unique trajectory for each the particle, the vortex ball in our case. However, an attempt to measure exact position of the ball along the trajectory together with its velocity fails. Namely, there is no way to measure simultaneously the complementary parameters, such as coordinate and velocity, what follows from the uncertainly principle [37].

Fig. 7 shows in lilac color Bohmian trajectories divergent from the slit grating. The probability density distribution is shown here in grey color ranging from white for p=0 to light grey for max p. Bundle of the Bohmian trajectories imitates a fluid flow through the obstacle, containing slits, relatively well. One can see that characteristic streamlets are formed in the flow, along which particles move. Such a vision of hydrodynamical behaviors of quantum systems is typical for many scientists since the formation of the quantum mechanics up to our days [38-42]. Principal moment is that the Schrödinger equation describes the expiration of the superfluid medium, which depend on the boundary conditions and other devices perturbing it (as, for example, the slit gratings, collimators and others). The vortex balls move along optimal directions of the flows – along the Bohmian paths.

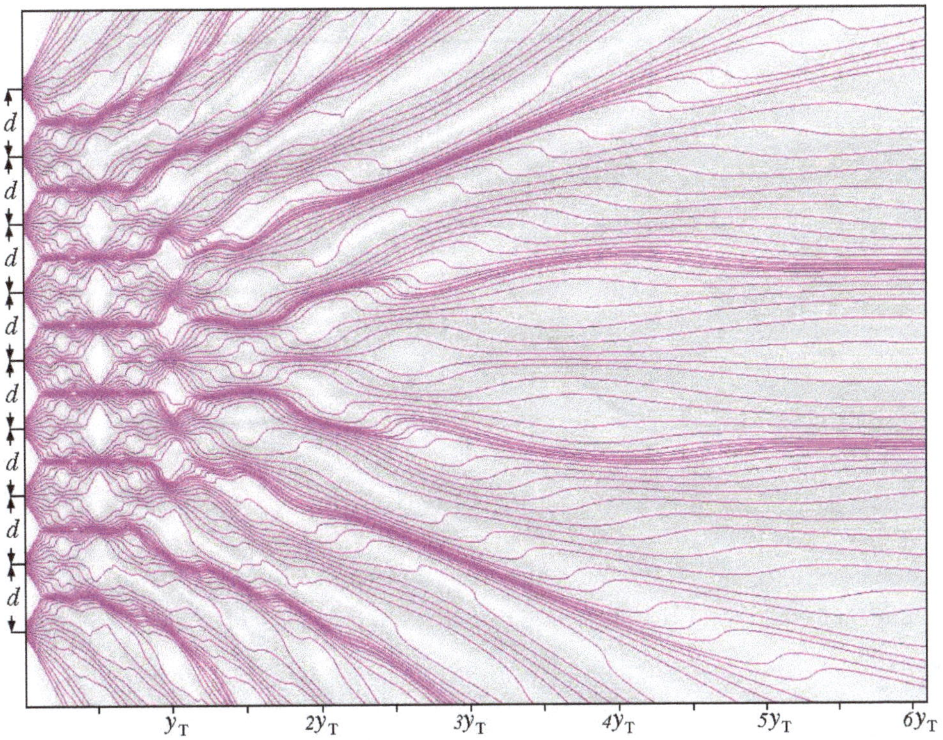

Figure 7. Interference pattern of the coherent flow of the fullerene molecules with the de Broglie wavelength $\lambda = 5$ pm within a zone $y \le 6y_T$ from the grating containing 9 slits. Lilac curves against the grey background represent the Bohmian trajectories.

4. Physical vacuum as a superfluid medium

The Schrödinger equation (34) describes a flow of the peculiar fluid that is the physical vacuum. The vacuum contains pairs of particle-antiparticles. The pair, in itself, is the Bose particle that stays at a temperature close to zero. In aggregate, the pairs make up Bose-Einstein condensate. It means that the vacuum represents a superfluid medium [43]. A 'fluidic' nature of the space itself is exhibited through this medium. Another name of such an 'ideal fluid' is the ether [29].

The physical vacuum is a strongly correlated system with dominating collective effects [44] and the viscosity equal to zero. Nearest analogue of such a medium is the superfluid helium [22], which will serve us as an example for further consideration of this medium. The vacuum is defined as a state with the lowest possible energy. We shall consider a simple vacuum consisting of electron-positron pairs. The pairs fluctuate within the first Bohr orbit having energy about $13.6 \cdot 2\,\mathrm{eV} \approx 27\,\mathrm{eV}$. Bohr radius of this orbit is $r_1 \approx 5.29 \cdot 10^{-11}$ m. These fluctuations occur about the center of their masses. The total mass of the pair, m_p, is equal to doubled mass of the electron, m. The charge of the pair is zero. The total spin of the pair is equal to 0. The angular momentum, L, is nonzero, however. For the first Bohr orbit $L = \hbar$. The velocity of rotating about this orbit is $L\ /(r_1 m) \approx 2.192 \cdot 10^6$ m/s. It means that there exist an elementary vortex. Ensemble of such vortices forms a vortex line.

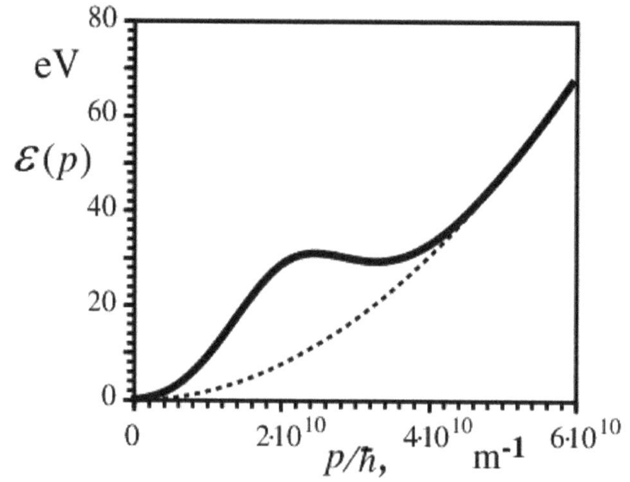

Figure 8. The dispersion relation ε vs. p. The dotted curve shows the non-relativistic square dispersion relation $\varepsilon \sim p^2$. The hump on the curve is a contribution of the roton component $p_R f(p-p_R)$, $p_R/\hbar \approx 1.89 \cdot 10^{10}$ m^{-1}, and $\sigma = 0.5p_R$.

We may evaluate the dispersion relation between the energy, $\varepsilon(p)=\hbar\omega$, and wave number, $p=\hbar k$, as it done in [45]. As follows from the Schrödinger equation (34) we have:

$$\varepsilon(p) = \frac{1}{2m_p}\left(p + p_R f(p-p_R)\right)^2. \tag{39}$$

Here $p_R = L/r_1 = m_p v_R$ is the momentum of the rotation. The function $f(p\text{-}p_R)$ is a form-factor relating to the electron-positron pairs rotating about the center of their mass m_p. The form-factor describes dispersion of the momentum p around p_R conditioned by fluctuations about the ground state with the lowest energy. The form-factor is similar to the Gaussian curve

$$f(p-p_R) = \exp\left\{-\frac{(p-p_R)^2}{2\sigma^2}\right\}. \tag{40}$$

Here σ is the variance of this form-factor. It is smaller or close to p_R. The dispersion relation (39) is shown in Fig. 8. The hump on the curve is due to the contribution of the rotating electron-positron pair about the center of their masses. These rotating objects are named rotons [45].

Rotons are ubiquitous in vacuum because of a huge availability of pairs of particle-antiparticle. The movement of the roton in the free space is described by the Schrödinger equation

$$\mathrm{i}\hbar\frac{\partial \Psi}{\partial t} = \frac{1}{2m_p}(-\mathrm{i}\hbar\nabla + m_p\vec{v}_R)^2\Psi - C\Psi. \tag{41}$$

The constant C determines an uncertain phase shift of the wave function, and most possible this phase relates to the chemical potential of a boson (the electron-positron pair) [45]. We shall not take into account contribution of this term in the dispersion diagram because of its smallness. As follows from the above consideration of the Navier-Stokes equation, Eq. (41) can be reduced to the Euler equation

$$\frac{\partial \vec{v}_R}{\partial t} + [\vec{\omega} \times \vec{v}_R] = -\frac{\nabla P}{m_p \rho}, \tag{42}$$

that describes a flow of the inviscid incompressible fluid under the pressure field P. One can see from here that the Coriolis force appears as a restoring force, forcing the displaced fluid particles to move in circles. The Coriolis force is the generating force of waves called inertial waves [45]. The Euler equation admits a stationary solution for uniform swirling flow under the pressure gradient along z.

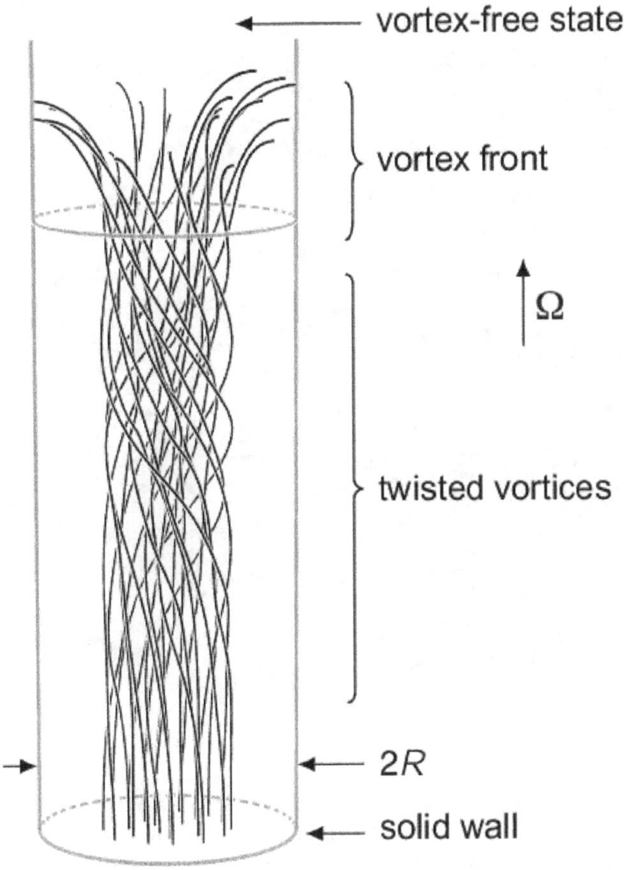

Figure 9. The formation of twisted vortex state [46]. The vortices have their propagating ends bent to the side wall of the rotating cylinder. As they expand upwards into the vortex-free state, the ends of the vortex lines rotate around the cylinder axis. The twist is nonuniform because boundary conditions allow it to unwind at the bottom solid wall. The figure gives a snapshot (at time $t = 25\Omega^{-1}$, where Ω is the angular velocity.) of a numerical simulation of 23 vortices initially generated near the bottom end (t=0). Courtesy kindly by Erkki Thuneberg.

Formation of the swirling flow, the twisted vortex state, has been studied in the superfluid ^{3}He-B [46]. These observations give us a possibility to suppose the existence of such phenomena in the physical vacuum. The twisted vortex states observed in the superfluid ^{3}He-B are closely related to the inertial waves in rotating classical fluids. The superfluid initially is at rest [46]. The vortices are nucleated at a bottom disk platform rotating with the angular velocity Ω about axis z. As the platform rotates they propagate upward by creating the twisted vortex state spontaneously, Fig. 9. The Coriolis forces take part in this twisting. The twisted vortices grow upward along the cylinder axis [47].

Analogous experiment with nucleating vortices can be realized when the lower disk A rotates in the vacuum, Fig. 10. In this case, the vortices are viewed as the dancing electron-positron pairs on the first Bohr orbit. As the vortices grow upward the spontaneous twisted vortex states arise. The latter by reaching upper fixed disk B can capture it into rotation.

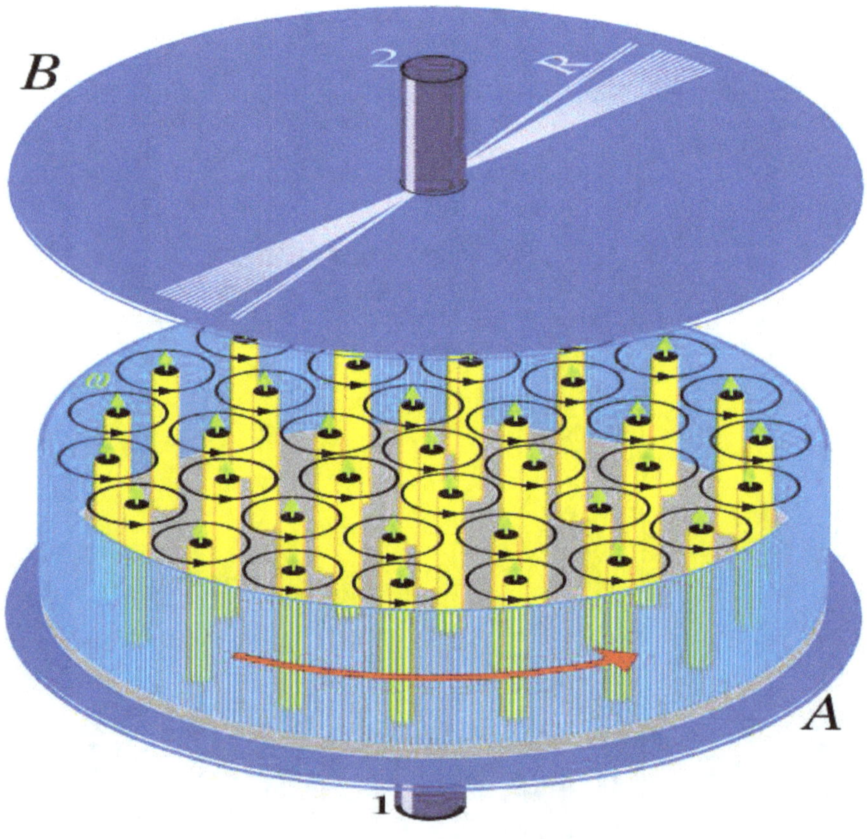

Figure 10. Rotation of the superfluid fluid is not uniform but takes place via a lattice of quantized vortices, whose cores (colored in yellow) are parallel to the axis of rotation [46, 47]. Green arrows are the vorticity ω. Small black arrows indicate the circulation of the velocity v_R around the cores. The vacuum is supported between two non-ferromagnetic disks, A and B, fixed on center shafts, **1** and **2**, of electric motors [48], see Fig. 11. Radiuses of the both disks are R=82.5 mm and distance between them can vary from 1 to 3 mm and more. The vortex bundle rotates rigidly with the disk A. As soon as the vortex bundle reaches the top disk B it begins rotation as well.

Pr. V. Samohvalov has shown through the experiment [48], that the vortex bundle induced by rotating the bottom non-ferromagnetic disk A leads to rotation of the upper fixed initially non-ferromagnetic disk B, Fig. 11. Both disks at room temperature have been placed in the container with technical vacuum at 0.02 Torr, The utmost number of the vortices that may be placed on the square of the disk A is $N_{\max}=(2\pi R^2)/(2\pi r_1^2)=2\bullet10^{18}$, where R=82.5 mm is the radius of the disk and $r_1\approx5.29\cdot10^{-11}$ m is the radius of the first Bohr orbit. Really, the number of the vortices situated on the square, N, is considerably smaller. It can be evaluated by multiplying $N_{\max}$ by a factor δ. This factor is equal to the ratio of the geometric mean of the velocities $v_R=\hbar/(r_1 m)\approx2.192\cdot10^6$ m/s and $V_D=R\Omega$ to their arithmetic mean. Here Ω is an angular rate of the disk A. So, we have

$$N = N_{\max}\frac{\sqrt{v_R\cdot V_D}}{v_R+V_D} = N_{\max}\sqrt{\frac{V_D}{v_R}} \approx 6\cdot10^{15} \qquad (43)$$

at the angular rate Ω=160 1/s [48] the disk velocity V_D=13.2 m/s. Now we can evaluate the kinetic energy of the vortex bundle induced by the rotating disk A. This kinetic energy is $E=N\cdot m_p v_R^2/2\approx$ 0.026 J. This energy is sufficient for transfer of the moment of force to the disk B. Measured in the experiment [48] the torque is about 0.01 N•m. So, the disk B can be captured by the twisted vortex.

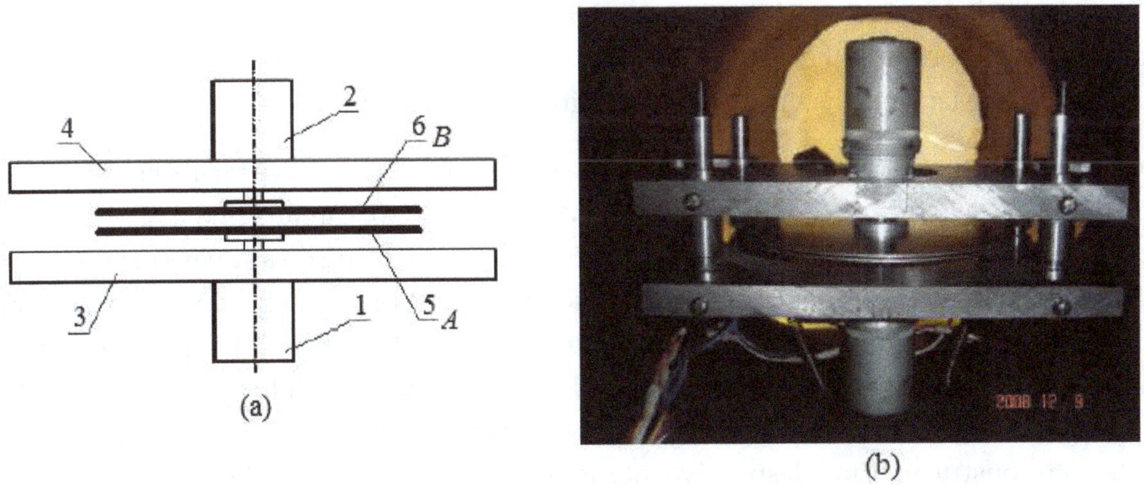

Figure 11. Basic diagram (a) and general view of the device (b) for researching mass dynamics effects [48]: 1 and 2 are shafts with mounted on them electric motors; 3 and 4 are steel plates with mounted on them electromagnetic brakes; 5 and 6 (the disks A and B, see Fig. 10) are disks rigidly fixed on flanges of the rotors of the electric motors. Courtesy kindly by Vladimir Samokhvalov.

The formation of the growing twisted vortices can be confirmed with attraction of modern methods of interference of light rays passing through the gap between the disks. Light traveling along two paths through the space between the disks undergoes a phase shift manifested in the interference pattern [29] as it was shown in the famous experiment of Aharonov and Bohm [49].

5. Conclusion

The Schrödinger equation is deduced from two equations, the continuity equation and the Navier-Stokes equation. At that, the latter contains slightly modified the gradient pressure term, namely, $\nabla P \to \rho_M \nabla (P/\rho_M) = \nabla P + P\nabla \ln(\rho_M)$. The extra term $P\nabla \ln(\rho_M)$ describes change of the pressure induced by change of the entropy $\ln(\rho_M)$ per length. In this case, the modified gradient pressure term can be reduced to the quantum potential through using the Fick's law. In the law we replace also the diffusion coefficient by the factor $\hbar/2m$, where $\hbar$ is the reduced Planck constant and m is mass of the particle.

We have shown that a vortex arising in a fluid can exist infinitely long if the viscosity undergoes periodic oscillations between positive and negative values. At that, the viscosity, in average on time, stays equal to zero. It can mean that the fluid is superfluid. In our case, the superfluid consists of pairs of particle-antiparticle representing the Bose-Einstein condensate.

As for the quantum reality, such a periodic regime can be interpreted as exchange of the energy quanta of the vortex with the vacuum through the zero-point vacuum fluctuations. In reality, these fluctuations are random, covering a wide range of frequencies from zero to infinity. Based on this observation we have assumed that the fluctuations of the vacuum ground state can support long-lived existence of vortex quantum objects. The core of such a vortex has nonzero radius inside of which the velocity tends to zero. In the center of the vortex, the velocity vanishes. The velocity reaches maximal values on boundary of the core, and then it decreases to zero as the distance to the vortex goes to infinity.

The experimental observations of the Couder's team [5, 11, 12] can have far-reaching ontological perspectives in regard of studying our universe. Really, we can imagine that our world is represented by myriad of baryonic and lepton "droplets" bouncing on a super-surface of some unknown dark matter. A layer that divides these "droplets", i.e., particles, and the dark matter is the superfluid vacuum medium. This medium, called also the ether [24], is populated by the particles of matter ("droplets"), which exist in it and move through it [29, 50, 51]. The particle traveling through this medium perturbs virtual particle-antiparticle pairs, which, in turn, create both constructive and destructive interference at the forefront of the particle [30]. Thus, the virtual pairs interfering each other provide an optimal, Bohmian, path for the particle.

Assume next, that the baryonic matter is similar, say, on "hydrophobic" fluid, whereas the dark matter, say, is similar to "hydrophilic" fluid. Then the baryonic matter will diverge each from other on cosmological scale owing to repulsive properties of the dark matter, like soap spots diverge on the water surface. Observe that this phenomenon exhibits itself through existence of the short-range repulsive gravitational force that maintains the incompatibility between the dark matter and the baryonic matter [52, 53]. At that, the dark matter stays invisible. One can imagine that the zero-point vacuum fluctuations are nothing as weak ripples on a surface of the dark matter.

Appendix A: Nelson's derivation of the Schrödinger equation

Nelson proclaim that the medium through which a particle moves contains myriad sub-particles that accomplish Brownian motions by colliding with each other chaotically. The Brownian motions is described by the Wiener process with the diffusion coefficient

$$\bar{\nu} = \frac{\hbar}{2m}. \tag{44}$$

Here m is mass of the particle and $\hbar = h/2\pi$ is the reduced Planck constant. Here we use ν with the upper bar in order to avoid confusion with the kinematic viscosity adopted in hydrodynamics. As seen this motion has a quantum nature [24] in contrast to the macroscopic Brownian motions where the diffusion coefficient has a view $\bar{\upsilon} = kT/m\beta^{-1}$; here k is Boltzmann constant, T is a temperature, and β^{-1} is the relaxation time.

Two equations are main in the article [24]. The position $\mathbf{x}$ (t) of the Brownian particle, being subjected either by external forces or by currents in the medium, can be written by two equivalent equations:

$$d\mathbf{x}(t) = \mathbf{b}(\mathbf{x}(t),t)dt + d\mathbf{w}(t), \tag{45}$$

$$d\mathbf{x}(t) = \mathbf{b}_*(\mathbf{x}(t),t)dt + d\mathbf{w}_*(t) \tag{46}$$

Here $w(t)$ and $w_*(t)$ are the Wiener processes, both have equivalent properties. Variables $\boldsymbol{b}$ and $\boldsymbol{b}_*$ are vector-valued forward and backward functions on space-time, respectively. In fact, they are the mean forward and mean backward measured quantities

$$\mathbf{b}(\mathbf{x}(t),t) = \lim_{\Delta t \to 0_+} E_t \frac{\mathbf{x}(t+\Delta t) - \mathbf{x}(t)}{\Delta t}, \tag{47}$$

$$\mathbf{b}_*(\mathbf{x}(t),t) = \lim_{\Delta t \to 0_+} E_t \frac{\mathbf{x}(t) - \mathbf{x}(t-\Delta t)}{\Delta t}. \tag{48}$$

Here E_t denotes the conditional expectation (average) given the state of the system at time t, and 0_+means that Δt tends to 0 through positive values. Thus $\boldsymbol{b}(x(t),\ t)$ and $\boldsymbol{b}_*(x(t),\ t)$ are again stochastic variables [54, 55]. It is instructive to compare calculus (46) and (47) with classical calculations of infinitesimal small increments

$$v(t) = \lim_{\Delta t \to 0_+} \frac{x(t+\Delta t) - x(t)}{\Delta t} = \lim_{\Delta t \to 0_+} \frac{x(t) - x(t-\Delta t)}{\Delta t}. \tag{49}$$

One can see that these calculations are symmetrical with respect to the time arrow, whereas the calculations (46) and (47) are not, in general (see below).

It should be noted that $b(x(t), t)$ and $b_*(x(t), t)$ are not real velocities. The real current velocity of the particle is calculated as

$$\vec{v}(t) = \frac{1}{2}\big(\mathbf{b}(\mathbf{x}(t),t) + \mathbf{b}_*(\mathbf{x}(t),t)\big). \tag{50}$$

There is a one more velocity, which is represented via difference of $b(x(t), t)$ and $b_*(x(t), t)$:

$$\vec{u}(t) = \frac{1}{2}\big(\mathbf{b}(\mathbf{x}(t),t) - \mathbf{b}_*(\mathbf{x}(t),t)\big). \tag{51}$$

According to Einstein's theory of Brownian motion, $\vec{u}(t)$ is the velocity acquired by a Brownian particle, in equilibrium with respect to an external force, to balance the osmotic force [19]. For this reason, this velocity is named the osmotic velocity. It can be expressed in the following form

$$\vec{u}(t) = \bar{\nu}\nabla\big(\ln(\rho(t))\big) = \frac{\hbar}{m}\frac{\nabla R(t)}{R(t)}, \tag{52}$$

where $\rho(t)$ is the probability density of $\mathbf{x}(t)$ and $R(t)=\rho(t)^{1/2}$ is the probability density amplitude. The current velocity, in turn, is expressed through gradient of a scalar field S called the action

$$\vec{v}(t) = -\frac{\hbar}{m}\nabla S(t). \tag{53}$$

The both equations, (44) and (45), introduced above are important for derivation of the Schrödinger equation. The derivation of the equation is provided by the use of the wave function presented in the polar form

$$\Psi = R\exp\{\,\mathrm{i}S/\hbar\,\}, \tag{54}$$

by replacing the velocities $\vec{v}(t)$ and $\vec{u}(t)$ in the initial equations. It should be noted that Nelson departs from two equations describing directed the forward and backward Brownian motions which are written down for real-valued functions. In order to come to the Schrödinger equation he has used a complex-valued wave function exp{R+iS} instead of the generally accepted $R\exp\{iS/\hbar\}$. Obviously, this discrepancy are eliminated by replacing $\exp\{R\}\to R$.

Observe that the wave function represented in the polar form (53) is used for getting equations underlying the Bohmian mechanics [27]. These two equations are the continuity equation and the Hamilton-Jacobi equation containing an extra term known as the Bohmian quantum potential. The quantum potential has the following view:

$$Q=-\frac{\hbar^2}{2m}\frac{\nabla^2 R}{R}=-\frac{\hbar}{2}\frac{(\nabla R\vec{u})}{R}=-\frac{m}{2}u^2-\frac{\hbar}{2}(\nabla\vec{u}). \quad (55)$$

One can see that the quantum potential depends only on the osmotic velocity, which is expressed through difference of the forward and backward averaged quantities (46) and (47). These forward and backward quantities can be interpreted as uncompensated flows through a "semipermeable membrane" which represents an instant dividing the past and the future. Following to Licata and Fiscaletti, who have shown that the quantum potential has relation to the Bell length indicating a non-local correlation [28], one can add that the non-local correlation exists also between the past and the future. E. Nelson as one can see has considered a particle motion through the ether populated by sub-particles experiencing accidental collisions with each other. The Brownian motions of the sub-particles submits to the Wiener process with the diffusion coefficient ν proportional to the Plank constant as shown in Eq. (43). The ether behaves itself as a free-friction fluid.

Acknowledgements

The author thanks Mike Cavedon for useful and valuable remarks and offers. The author thanks also Miss Pipa (quantum portal administrator) for preparing a program drawing Fig. 7.

Author details

V.I. Sbitnev[1,2]

1 St. Petersburg B.P. Konstantinov Nuclear Physics Institute, NRC Kurchatov Institute, Gatchina, Leningrad district, Russia

2 Department of Electrical Engineering and Computer Sciences, University of California, Berkeley, USA

References

[1] Hartle JB. What Connects Different Interpretations of Quantum Mechanics? In: Elitzur A., Dolev S., Kolenda N. (eds.) Quo Vadis Quantum Mechanics? Springer: Berlin Heidelberg New York; 2005. p72-82.

[2] Juffmann T, Truppe S, Geyer P, Major AG, Deachapunya S, Ulbricht H, Arndt M. Wave and particle in molecular interference lithography. Phys. Rev. Lett. 2009; 103: 263601.

[3] Brukner Č, Zeilinger A. Quantum Physics as a Science of Information. In: Elitzur A., Dolev S., Kolenda N. (eds.) Quo Vadis Quantum Mechanics? Springer: Berlin Heidelberg New York; 2005. p46-61.

[4] De Broglie L. Interpretation of quantum mechanics by the double solution theory. Annales de la Fondation Louis de Broglie. 1987; 12(4): 1-22.

[5] Couder Y, Fort E. Single-Particle Diffraction and Interference at a Macroscopic Scale. Phys. Rev. Lett. 2006; 97: 154101.

[6] Motte A. Translation of Newton's Principia (1687). Axioms or Laws of Motion. Published by Daniel Adee, 45 Liberty str., N. Y.; 1846.

[7] Frisch U. Translation of Leonhard Euler's: general principles of the motion of fluids. e-print: http://arxiv.org/abs/0802.2383 (accessed 17 Feb 2008).

[8] Sbitnev VI. From the Newton's laws to motions of the fluid and superfluid vacuum: vortex tubes, rings, and others. e-print: http://arxiv.org/abs/1403.3900 (accessed 20 Jun 2014).

[9] Landau LD., Lifshitz EM. Fluid mechanics. Pergamon Press; 1987.

[10] Kundu P, Cohen I. Fluid Mechanics. Academic Press; 2002.

[11] Couder Y, Fort E, Gautier C.-H, Boudaoud A. From bouncing to floating drops: non-coalescence of drops on a fluid bath. Phys. Rev. Lett. 2005; 94: 177801.

[12] Eddi A, Sultan E, Moukhtar J, Fort E, Rossi M, Couder Y. Information stored in Faraday waves: the origin of a path memory. J. Fluid Mech. 2011; 674: 433-463.

[13] Grössing G. On the thermodynamic origin of the quantum potential. Physica A, 2009; 388: 811-823.

[14] Grössing G. Sub-Quantum Thermodynamics as a Basis of Emergent Quantum Mechanics. Entropy, 2010; 12: 1975-2044. doi:10.3390/e12091975

[15] Grössing G., Mesa Pascasio J., Schwabl H. A classical explanation of quantization, Found. Phys. 2011: 41, 1437-1453.

[16] Grössing G. Emergence of Quantum Mechanics from a Sub-Quantum Statistical Mechanics. e-print: http://arxiv.org/abs/1304.3719 (accessed 12 Apr 2013).

[17] Harris DM, Bush WM. Droplets walking in a rotating frame: from quantized orbits to multimodal statistics. J. Fluid Mech. 2014; 739: 444-464.

[18] Oza AU, Harris DM, Rosales RR, Bush WM. Pilot-wave dynamics in a rotating frame: on the emergence of orbital quantization. J. Fluid Mech. 2014; 744: 404-429.

[19] Brady R. Anderson R. Why bouncing droplets are a pretty good model of quantum mechanics? e-print: http://arxiv.org/abs/1401.4356 (accessed 16 Jan 2014).

[20] Vervoort L. No-Go Theorems face Fluid-Dynamical Theories for Quantum Mechanics. e-print: http://arxiv.org/abs/1406.0901 (accessed 16 Jun 2014).

[21] Wu Jie-Zhi, Ma Hui-Yang, Zhou Ming-De. Vorticity and Vortex Dynamics. Springer-Verlag: Berlin Heidelberg; 2006.

[22] Volovik GE. The Universe in a Helium Droplet. Clarendon Press: Oxford; 2003.

[23] Tropical cyclone. http://en.wikipedia.org/wiki/Tropical_cyclone

[24] Nelson E. Derivation of the Schrödinger Equation from Newtonian Mechanics. Phys. Rev. 1966; 150(4): 1079-1085.

[25] Hestenes D. The Zitterbewengung interpretation of quantum mechanics. Found. Physics. 1990; 20(10): 1213-1232.

[26] Dorbolo S, Terwagne D, Vandewalle N, Gilet T. Resonant and rolling droplet. New Journal of Physics. 2008; 10: 113021. http://iopscience.iop.org/1367-2630/10/11/113021

[27] Sbitnev VI. Bohmian Trajectories and the Path Integral Paradigm – Complexified Lagrangian Mechanics. In: Pahlavani MR (ed.) Theoretical Concepts of Quantum Mechanics. Rijeka: InTech; 2012. p313-34; doi:10.5772/33064

[28] Licata I, Fiscaletti D. Bell Length as Mutual Information in Quantum Interference. Axioms 2014; (3): 153-165. doi:10.3390/axioms3020153

[29] Martins AA. Fluidic Electrodynamics: On parallels between electromagnetic and fluidic inertia. e-print: http://arxiv.org/abs/1202.4611 (accessed 21 Feb 2012).

[30] Feynman RP, Hibbs A. Quantum Mechanics and Path Integrals. McGraw Hill: N. Y.; 1965.

[31] Derbes D. Feynman's derivation of the Schrödinger equation. Am. J. Phys. 1996; 64(7) 881-884.

[32] Sbitnev VI. Matter waves in ihe Talbot-Lau interferometry. e-print: http://arxiv.org/abs/1005.0890 (accessed 17 Sep 2010).

[33] Talbot HF. Facts Relating to Optical Science. Philos. Mag. 1836; 9: 401-407.

[34] Berry MV, Klein S. Integer, fractional and fractal Talbot effects, Journal of Modern Optics. 1996; 43(10): 2139-2164.

[35] Berry M, Marzoli I, Schleich W. Quantum carpets, carpets of light. Phys. World. 2001: 6; 39-44.

[36] Sanz AS, Miret-Artés S. A causal look into the quantum Talbot effect, J. Chem. Phys. 2007; 126: 234106. http://arxiv.org/abs/quant-ph/0702224.

[37] Sbitnev VI. Generalized Path Integral Technique: Nanoparticles Incident on a Slit Grating, Matter Wave Interference. In: Bracken P (ed.) Advances in Quantum Mechanics. Rijeka: InTech; 2013. p183-211. doi:10.5772/53471

[38] Madelung E. Quantumtheorie in hydrodynamische form. Zts. f. Phys. 1926; 40: 322-326.

[39] Bohm D, Vigier JP. Model of the causal interpretation of quantum theory in terms of a fluid with irregular fluctuations. Phys. Rev. 1954; 96: 208-216.

[40] Wyatt RE. Quantum dynamics with trajectories: Introduction to quantum hydrodynamics. Springer: 2005.

[41] Jüngel A, Milišić J-P. Quantum Navier–Stokes equations. Progress in Industrial Mathematics at ECMI 2010. Mathematics in Industry. 2012; 17: p427-439.

[42] Guo Y, Bühler O. Wave–vortex interactions in the nonlinear Schrödinger equation. Physics of Fluids. 2014; 26: 027105.

[43] Sinha KP, Sivaram C, Sudarshan ECG. Aether as a Superfluid State of Particle-Antiparticle Pairs. Found. Phys. 1976; 6(1): 65-70.

[44] Roberts PH, Berloff NG. The nonlinear Schrödinger equation as a model of superfluidity. In: Barenghi CF, Donnelly RJ, Vinen WF. (eds.) Quantized vortex dynamics and superfluid turbulence. LNP 571. 2001: 235-257.

[45] Le Gal P. Waves and Instabilities in Rotating and Stratified Flows. In: Klapp J, Medina A, Cros A, Vargas CA. (eds.) Fluid Dynamics in Physics, Engineering and Enviromental Applications. Springer-Verlag; 2013. p25-40.

[46] Eltsov VB, Finne AP, Hänninen R, Kopu J, Krusius M, Tsubota M, Thuneberg EV. Twisted vortex states. Phys. Rev. Lett. 2006; 96: 215302; http://xxx.lanl.gov/abs/cond-mat/0602667.

[47] Lounasmaa OV, Thuneberg E. Vortices in rotating superuid ^{3}He. PNAS, 1999; 96(14): 7760-7767.

[48] Samokhvalov VN. Nonelectromagnetic force interaction by rotating masses in vacuum. Int. J. Unconventional Science. 2013; 1(1): 6-19. http://www.unconv-science.org/n1/samokhvalov/

[49] Aharonov Y, Bohm D. Significance of electromagnetic potentials in the quantum theory. Phys. Rev. 1959; 115(3): 485-491.

[50] Pinheiro MJ, Büker M. An extended dynamical equation of motion, phase dependency and inertial back reaction. e-print: http://arxiv.org/abs/1208.3458 (accessed 16 Aug 2012).

[51] Saravani M, Afshordi N, Mann RB. Empty Black Holes, Firewalls, and the Origin of Bekenstein-Hawking Entropy. e-print: http://arxiv.org/abs/1212.4176 (accessed 2 Dec 2013).

[52] Famaey B, McGaugh S. Modified Newtonian Dynamics (MOND): Observational Phenomenology and Relativistic Extensions. Living Review Relativity. 2012; 15: 10.

[53] Chung Ding-Yu. Galaxy Evolution by the Incompatibility between Dark Matter and Baryonic Matter. Int. J. Astronomy and Astrophysics. 2014; 4: 374-383.

[54] Nelson E. Dynamical theories of Brownian motion. Princeton. New Jersey: Princeton Univ. Press; 1967.

[55] Nelson E. Quantum fluctuations. Princeton Series in Physics. Princeton, New Jersey: Princeton Univ. Press; 1985.

5

Photons and Signals in the Age of Information

Cynthia Kolb Whitney

1. Introduction

The history of Physics contains some inexplicable mysteries, and one of them is this: back in the early 20th century, Einstein was working on the idea of the 'Photon' and on the idea of the 'Signal' at essentially the same time [1,2], but he did not relate them to each other. They seem to have arisen totally separate in his mind, and they led to totally separate subsequent developments. The Photon played a central role in the development of Quantum Mechanics (QM), and the Signal played the central role in the development of Special Relativity Theory (SRT).

QM and SRT are now the two great pillars of early 20th century Physics, but they seem to be in conflict over the issue of communication. In QM, Schrödinger's Equation is basically the Fourier transform (just a restatement in terms of different variables) of a statement from Classical Mechanics: Total energy = kinetic energy + potential energy. This statement has no signal propagation speed involved in it. So QM appears to allow instantaneous communication over arbitrary distances. But in SRT, Einstein's Second Postulate limits *all* communication to light speed, c.

Since QM and SRT conflict so dramatically on the issue of communication, at least one of them must, in some sense, be wrong. So we are left unsure about what to believe, or take as a foundation for future development.

Can we find out anything decisive from experiments or observations?

The foundation for QM is usually called the Quantum Hypothesis, and not the Quantum Postulate, because the granularity aspect of QM is experimentally testable in various ways. So far, the granularity aspect of QM never seems to fail. This fact stands in favor of QM. But many people still do not believe in the instantaneous communication aspect of QM that is manifest in Schrödinger's Equation.

The foundation for SRT can be called the Light-Speed Postulate, but not the Light-Speed Hypothesis, because light speed really is *not* experimentally testable, since a test would involve at least two different spacetime points, and the correlation of data from two different spacetime points would involve the Light-Speed Postulate itself. Despite numerous claims to the contrary, SRT has *not* been tested in a way that actually could have falsified it, and only very indirect testing appears even feasible for SRT.

Einstein's General Relativity Theory (GRT) flows from SRT, and GRT is testable, at least observationally, although not experimentally. But the new hypotheses that it offers for observational test are few in number, and great in technical difficulty. So even indirect testing of SRT through GRT does not look very promising.

An altogether different approach therefore seems needed: instead of demanding experiments or observations, we should be reviewing the founding ideas themselves, in light of new insights gathered in the intervening century. Evidently, at least one of the founding ideas, and possibly both of them, need to be updated, or else retired and replaced. This paper aims to identify possible update(s)/replacement(s) that may help.

Some of the needed insights come from engineering practice, rather than from theoretical physics. In the mid 20th century there was a flowering of Information Theory (IT), first in connection with wartime code breaking and code making, and then in connection with the post-war communication industry. All of that development led in turn to our modern computation industry, and our current 'Age of Information'.

IT uses the concept of Entropy, taken from classical Thermodynamics, and with the application of a minus sign, provides a quantitative mathematical measure for Information. This measure can be used in support of all sorts of engineering concept analyses and design decisions, *etc.* A convenient reference about the IT concepts and their general applications is Leon Brillouin's wonderful little book *Science and Information Theory* [3]. Flores Gallegos [4] discusses some particular applications in QM.

Viewed from our vantage point here in the early 21st century, IT actually provides a clear disqualifier for Einstein's Second Postulate. The problem is this: the Second Postulate is based on the behavior of a classical infinite plane wave, and an infinite plane wave cannot convey any information whatsoever!

The reason for this perhaps startling assertion is that an infinite plane wave is to electromagnetic communication what a steady hum is to auditory communication: background at best. There is no music in the monotonous hum, and there is no message in the infinite plane wave.

Information requires structure: amplitude modulation, or frequency modulation, or on-off switching. An infinite plane wave does not have any such structure. Because of this deficit, we certainly need a new Signal model as the foundation for an updated SRT.

Possibly we also need a better Photon model at the foundation of QM. We do not actually have a detailed and universally accepted model for photons in QM. Attention has focused more on material systems, which are said to change state, and in so doing, emit or absorb photons that carry packets of energy and angular momentum. As for electric and magnetic fields, what we have is the notion from Quantum Electrodynamics of a 'virtual photon'. This terminology reveals the desirability of having some unified, photon-like, approach for both Coulomb-Ampère and radiation fields, but it does not actually provide details.

Without detail to argue against, we are apparently free to develop a Photon model *de novo*, in a way that serves not only as the Photon model for QM, but also provides the more realistic Signal model needed for an updated SRT.

The new Photon/Signal model need *not* involve yet another new Postulate. Remember what Euclid taught the world through his Geometry: use no more Postulates than absolutely necessary. The reason is that unnecessary Postulates can conflict with other Postulates already in place, and so lead to Paradoxes.

In SRT, we do indeed have many Paradoxes, involving rods, clocks, trains, lightening strikes, snakes, barns, twins, and so on, and on. A prime suspect for their root cause is the unnecessary Second Postulate. In QM, Schrödinger's Equation seemingly came full-formed from heaven, and to that extent was also a Postulate, and indeed one that conflicted with Einstein's Second Postulate. The mysterious feel of quantum duality, for example, may suggest a possible QM Paradox yet to be fully articulated. If so, the new Photon/Signal model can offer a candidate approach to solve the problem.

The Photon/Signal model is just very familiar, old-fashioned mathematics: **1)** partial differential equations (Maxwell's first order coupled field equations), **2)** their family of solutions (starting with Gaussian pulses, and generating by differentiations the higher-and-higher order Hermite polynomials multiplying the Gaussian pulses), and **3)** boundary conditions (no backflow of energy behind the source, no overflow of energy beyond the receiver). This formulation is enough to determine the particular solution that fits any particular problem.

One general rule in Science is this: when a venerable and well-tested mathematical approach exists, try it **first**, before abandoning it in favor of a new approach. The irony is that Einstein chose not to apply usual the mathematical approach, and instead to introduce his additional Postulate. But he did it so long ago that, despite the long list of Paradoxes generated by SRT, his Second Postulate has itself become 'venerable', and therefore nearly impossible to unseat.

But with SRT updated with the new Photon/Signal model, based entirely on the old-fashioned mathematical approach, QM no longer needs to conflict with SRT. Atoms can have solution states in which energy loss by radiation is countered by an energy gain mechanism newly identified with the updated SRT. It is no longer necessary to postulate the Schrödinger equation just to prevent atomic death by loss of orbit energy to far-field radiation.

With n new starting point for QM, there can be a new development for QM. Some type of Quantum Gravity has long been sought, but with GRT being founded in SRT, with *c*-speed-only communication, and with QM being founded in Classical Physics, with instantaneous communication, that goal has been hard to reach. But the combination of an updated SRT and traditional Statistical Mechanics (SM) offers a way forward.

The new protocol for treating gravity is this: **1)** First notice that gravitational attraction formally resembles a statistical residue from magnetic interactions between elements of charge-neutral matter that are carrying electrical currents, 'current elements' for short. Current elements were well described by Ampère before Maxwell ever came along. **2)** Then apply SM to pairs of current elements.

QM also connects to modern technology problems. Chemistry is wonderfully rich application area for QM. But current-day Quantum Chemistry (QC) is not very easy to use, mainly because of heavy computation loads associated with numerical integrations. The new photon/signal model leads to a different approach for QC, one that is algebraic, rather than integral, in character. This paper includes some recent results from the application of Algebraic Chemistry (AC) to research on the chemistry of water; namely, the form of water known as 'EZ water', because it excludes positive ions.

Finally, QM may connect to Elementary Particle Physics in a manner yet to be fully developed. Just as the myriad compounds in Chemistry arises from not-very-many chemical elements, some significant part of the myriad of currently understood 'elementary' particles may arise from just two of them: the electron and the positron.

The suggestion for this idea lies in Chemistry's Periodic Table (PT). The way that the electron spin states fill up with increasing nuclear charge suggests the existence of not only electron pairs that have opposing spins, but also electron rings that have aligned spins. Electron rings involve up to three, five, even seven, electrons. Basically, electrons within atoms form, not only opposite-spin couples, but also same-spin teams. Since electron rings apparently do occur in atoms, they may well also occur separated from atomic nuclei, and therefore looking like exotic elementary particles. And the same may be said of positrons. Thus we have a rich array of possibilities yet to explore.

2. The photon / signal model

The first part of the Photon/Signal model consists of the governing partial differential equations. These are Maxwell's four first-order coupled field equations. Jackson [5] gives Maxwell's equations in modern notation and Gaussian units as:

$$\nabla \bullet \mathbf{B} = 0, \nabla \bullet \mathbf{D} = 4\pi\rho, \nabla \times \mathbf{E} + \frac{1}{c}\partial \mathbf{B} / \partial t = 0, \nabla \times \mathbf{H} - \frac{1}{c}\partial \mathbf{D} / \partial t = \frac{4\pi}{c}\mathbf{J}. \tag{1}$$

Here **B** is magnetic field and **E** is electric field. The constant $1/c = \sqrt{\varepsilon_0 \mu_0}$, where ε_0 is electric permittivity and μ_0 is magnetic permeability. In free space, $\mathbf{D}=\varepsilon_0\mathbf{E}$, $\mathbf{H}=\mathbf{B}/\mu_0$, and charge density ρ and current density **J** are zero. Free space will be the case of interest henceforth in this paper.

The second part of the Photon/Signal model is the family of suitable finite-energy solutions. This is developed as follows. Let the word 'pulse' be the short description for a field profile that is rounded on top and sloping down on the sides, and fading gradually to zero; for example, a Gaussian function.

The mechanism for the waveform development is that the spatial derivatives applied in Maxwell's equations change the original Gaussian pulse into successively longer wavelets consisting of successively higher order Hermite polynomials multiplying the original Gaussian. As a result, the energy in the wavelet gets more and more spread out along the propagation path.

Observe that waveform development is inexorable, just like ever-growing entropy in Thermodynamics. This is, I believe, where Entropy really enters into Physics. That is, Maxwell's first order coupled field equations are what give to Physics its obvious Arrow of Time. In citing electromagnetism as the cause for irreversibility, this idea follows Bentwich [6]. (It does *not*, however, give any hope for *reversing* anything.)

Let the spatial variable argument for the pulse be x. Let the direction of the initial pulse be y. Figure 1 illustrates this Gaussian pulse, along with a snapshot showing how it evolves over one complete cycle through Maxwell's first-order coupled field equations. Series 1 is the input pulse, and Series 2 is the waveform developed from it. A more complicated Figure was given in Whitney [7]. This simplified Figure 1 focuses on just the one issue: waveform development. The wavelet is shown bold because it has not been given sufficient attention before.

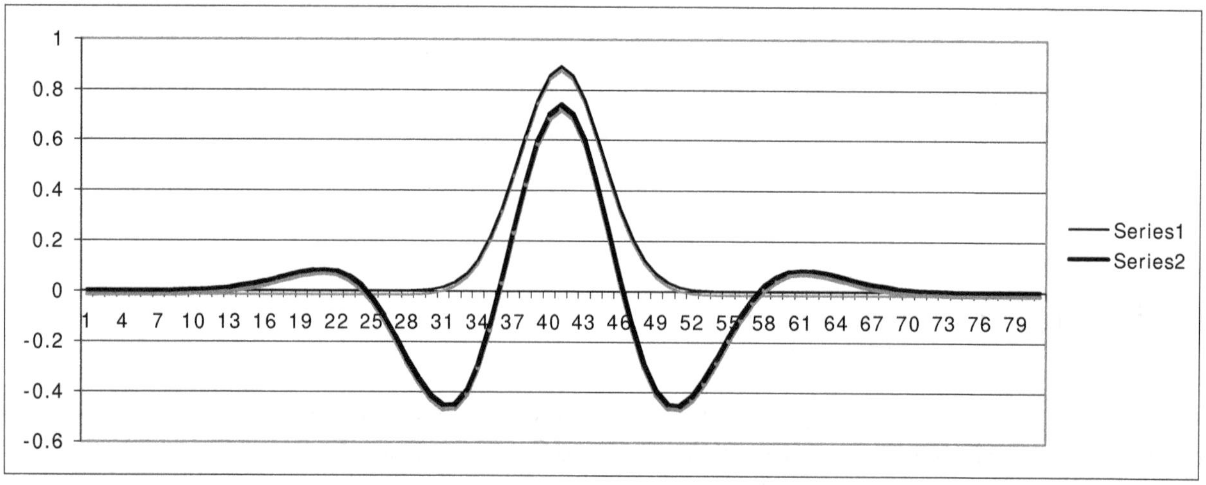

Figure 1. Illustration of waveform development.

Now, to solve the stated propagation problem, one can pose a primary pair of pulses in **E** and **B** to guarantee travel, and, for more realism, add a second such pair, offset a quarter cycle in time and perpendicular in space, to model circular polarization, like a real physical photon exhibits.

The third part of the Photon/Signal model is the pair of propagation boundary conditions: no backflow of energy behind the source, and no overflow of energy beyond the receiver. To guarantee these boundary conditions, one can demand **E**=0 at these boundaries.

One can fulfill the required zero **E** fields by matching the signal leaving the source toward the receiver with **1)** another fictitious signal going from the source in the opposite direction, so as to make the **E** field zero at the source, and **2)** another fictitious signal approaching the receiver from the opposite direction, so as to make the **E** field zero at the receiver. One can even continue this boundary-fixing process, to clean up each tiny new departure from zero **E** at a boundary that each additional fictitious signal creates at the end of the path opposite to the end that it is meant to correct.

Carried to an infinite sum of corrections upon corrections, this is certainly a very complicated picture about electromagnetic fields. How then can we extract from it some simple statement about propagation speed? We have all been trained to just say c. "But relative to what?" we might ask. If we knew the source and receiver were stationary relative to each other, we could be sure that $c = c_{\text{relative to receiver}}$, consistent with Einstein's Second Postulate. Otherwise, we would have to say more.

If we would focus attention to moments when the bulk of the energy is very near the receiver, we could be fairly confident that $c = c_{\text{relative to receiver}}$, again consistent with Einstein's Second Postulate. But if we would focus attention to moments when the bulk of the energy is still very near the source, moments when the receiver is nothing more than a distant phantom, we would be hard pressed to argue against the proposition that $c = c_{\text{relative to source}}$. (This would be consistent with the 1908 Ritz Proposal [8], which was much investigated in the early to mid 20th century as a candidate alternative to Einstein's Second Postulate, but was ultimately rejected.)

If we would need to characterize an entire propagation scenario, we would have to go even further, and consider *all* moments along the way. We would have neither the Einstein Second Postulate, nor the Ritz Proposal, but rather something else more complicated. It certainly must conform to Einstein's Second Postulate late in the scenario, when the bulk of the energy is near the receiver. But early in the scenario, when the bulk of the energy is still near the source, it must conform to Ritz's Proposal. And in between, it must represent in some mathematically appropriate way an idea not previously considered: a transition from one reference for c to the other.

Here is one way to formulate the problem. Let variable x represent distance along the propagation path, and let variable t represent time into the propagation process. At any point x, t there are fields $\mathbf{E}(x, t)$ and $\mathbf{B}(x, t)$ with magnitudes $E(x, t)$ and $B(x, t)$, and from them a local energy density:

$$S(x,t) = \left[E^2(x,t) + B^2(x,t) \right] / 2. \tag{2}$$

One way to characterize the propagation is with path integrals. [9-13] Consider the following ratio of two path integrals:

$$r(t) = \int_{\text{path}} x S(x,t) dx \Big/ \int_{\text{path}} S(x,t) dx. \tag{3}$$

This ratio provides a function that begins at the source, and ends at the receiver, and at the temporal midpoint of the scenario, gives equal weight to both the source and the receiver. This behavior captures the proposed transition of the reference to which the light speed c is assumed relative.

3. Using the changing reference for *c* changes the results

The concept of the changing reference for *c* to be relative to appears relevant for an important scenario that was considered even before Einstein arrived on the scene. The scenario involves the potentials and fields created by rapidly moving sources, and it was addressed starting in the late nineteenth century and very early twentieth centuries. Researchers then made the same Assumption that Einstein later made his Second Postulate in founding SRT, but they were not attentive enough to see that there indeed *was* an Assumption, and to call it out as his Second Postulate. So Einstein is to be commended for calling attention to this Assumption.

The sources generally cited for this early (1898 to 1901) problem are A. Liénard [14] and E. Wiechert [15]. Although they worked at about the same time, they worked separately. They got the same results, as did all contemporary and subsequent investigators, because all persons working from then up until now have used the same input Assumption; namely, that the speed of light is always *c* with respect to the receiver of the light.

The Liénard- Wiechert results are given in [5], and in every other standard EM book. They are displayed in [7], and that short passage is quoted again here, for review and subsequent further discussion:

"The standard scalar and vector potentials are:

$$\Phi(\mathbf{r},t)=e\left[1/\kappa R\right]_{\text{retarded}} \text{ and } \mathbf{A}(\mathbf{r},t)=e\left[\boldsymbol{\beta}/\kappa R\right]_{\text{retarded}}, \tag{4}$$

"where $\kappa=1-\mathbf{n}\cdot\boldsymbol{\beta}$, $\boldsymbol{\beta}$ is source velocity normalized by light speed *c*, and $\mathbf{n}=\mathbf{R}/R$ (a unit vector), and $\mathbf{R}=\mathbf{r}_{\text{source}}(t-R/c)-\mathbf{r}_{\text{receiver}}(t)$ (an implicit definition for the terminology 'retarded').

"The LW fields obtained from those potentials are then:

$$\begin{aligned}\mathbf{E}(\mathbf{r},t)&=e\left\{\frac{1}{\kappa^3R^2}(\mathbf{n}-\boldsymbol{\beta})(1-\beta^2)+\frac{1}{c\kappa^3R}\mathbf{n}\times\left[(\mathbf{n}-\boldsymbol{\beta})\times(d\boldsymbol{\beta}/dt)\right]\right\}_{\text{retarded}},\\ \mathbf{B}(\mathbf{r},t)&=\mathbf{n}_{\text{retarded}}\times\mathbf{E}(\mathbf{r},t).\end{aligned} \tag{5}$$

"The $1/R$ fields are radiation fields, and they make a Poynting vector (energy flow per unit area per unit time) that lies along $\mathbf{n}_{\text{retarded}}$:

$$\begin{aligned}\mathbf{P}&=\frac{c}{4\pi}\mathbf{E}_{\text{radiative}}\times\mathbf{B}_{\text{radiative}}\\ &=\frac{c}{4\pi}\mathbf{E}_{\text{radiative}}\times\left(\mathbf{n}_{\text{retarded}}\times\mathbf{E}_{\text{radiative}}\right)=\frac{c}{4\pi}\left(E_{\text{radiative}}\right)^2\mathbf{n}_{\text{retarded}}\ .\end{aligned} \tag{6}$$

"The $1/R^2$ fields are Coulomb-Ampère fields, and the Coulomb field

$$\mathbf{E}(\mathbf{r},t) = e\left\{(\mathbf{n}-\boldsymbol{\beta})(1-\beta^2)/\kappa^3R^2\right\}_{\text{retarded}}. \tag{7}$$

"does *not* lie along $\mathbf{n}_{\text{retarded}}$ as one might initially expect; instead, it lies along $(\mathbf{n}\text{-}\boldsymbol{\beta})_{\text{retarded}}$. Assume that $\boldsymbol{\beta}$ does not change much over the total field propagation time, in which case $(\mathbf{n}\text{-}\boldsymbol{\beta})_{\text{retarded}}$ is virtually indistinguishable from $\mathbf{n}_{\text{present}}$."

Thus the Coulomb attraction/repulsion and the radiation Poynting vector have distinctly different directions. This result does *not* look right physically. It looks as though advance information is being provided on one, but not the other, of two information channels.

Now consider the same problem using the new and more nuanced definition for the light speed reference. Observe that a line that connects the source position at the temporal midpoint of the scenario to the receiver position at the temporal midpoint of the scenario defines both the distance and the direction that the energy must travel in order to achieve the proposed transmission of the signal from the source to the receiver. This specification implies that we need potentials and fields to be, not *retarded*, but **half**-*retarded*. Now the potentials become:

$$\Phi(\mathbf{r},t) = e\left[1/\kappa R\right]_{\textbf{half}-\text{retarded}} \text{ and } \mathbf{A}(\mathbf{r},t) = e\left[\boldsymbol{\beta}/\kappa R\right]_{\textbf{half}-\text{retarded}}. \tag{8}$$

The fields become:

$$\mathbf{E}(\mathbf{r},t) = e\left\{\frac{(\mathbf{n}-\boldsymbol{\beta})(1-\beta^2)}{\kappa^3R^2} + \frac{\mathbf{n}}{c\kappa^3R}\times\left[(\mathbf{n}-\boldsymbol{\beta})\times d\boldsymbol{\beta}/dt\right]\right\}_{\textbf{half}-\text{retarded}} \text{ and } \mathbf{B}(\mathbf{r},t) = \mathbf{n}_{\textbf{half}-\text{retarded}}\times\mathbf{E}(\mathbf{r},t). \tag{9}$$

The Poynting vector becomes:

$$\begin{aligned}\mathbf{P} &= \frac{c}{4\pi}\mathbf{E}_{\text{radiative}}\times\mathbf{B}_{\text{radiative}} \\ &= \frac{c}{4\pi}\mathbf{E}_{\text{radiative}}\times\left(\mathbf{n}_{\textbf{half}-\text{retarded}}\times\mathbf{E}_{\text{radiative}}\right) = \frac{c}{4\pi}(E_{\text{radiative}})^2\mathbf{n}_{\textbf{half}-\text{retarded}}.\end{aligned} \tag{10}$$

The direction of the Coulomb field becomes:

$$(\mathbf{n}-\boldsymbol{\beta})_{\textbf{half}-\text{retarded}} \approx (\mathbf{n}_{\text{present}})_{\textbf{half}-\text{retarded}} \triangleq \mathbf{n}_{\textbf{half}-\text{retarded}}. \tag{11}$$

This direction is the same as the direction of the radiation Poynting vector. That is, the Coulomb field and the Poynting vector are now reconciled to the *same* direction, instead of conflicting with each other.

4. The corrected force direction means the hydrogen atom can survive classically

Many authors have expressed the opinion that really explaining the Hydrogen atom requires some presently-unknown short-range repulsive force between the electron and the proton; see, for example, Lokajicek, *et al* [16]. But given the results just presented, no mysterious new repulsive force is needed. With the direction of the Coulomb field being $\mathbf{n}_{\text{half-retarded}}$, there is a tiny tangential component of Coulomb force aligned with the orbit velocity. So there is a torque on the atom, and the torque pumps energy into the atom, and that process can work to balance the energy loss due to radiation.

That is to say: having a more nearly correct model for potentials and fields created by rapidly moving charges makes it possible to *explain* the immortality of the Hydrogen atom without first *postulating* the immortality of the Hydrogen atom; *i.e.*, postulating Schrödinger's equation.

That is to say: we need not postulate Schrödinger's equation; w can instead just carry out the old-fashioned math.

In my 2012 and 2013 papers, I listed just the pertinent results. Here is more detail and derivation:

Let the masses of the electron and the proton be m_e and m_p. Note that $m_e << m_p$, but m_p is *not* infinite.

Let the orbit radii of the electron and the proton be r_e and r_p. Note that $r_p << r_e$, but r_p is *not* zero.

Let the charges on the electron and the proton be $-e$ and $+e$.

The magnitude of the nominally attractive force within the atom is $F = e^2 / (r_e + r_p)^2$.

Let the orbit frequency be Ω. The orbit speed of the electron is $v_e = r_e \Omega$ and that of the proton is $v_p = r_p \Omega$.

The magnitude of the tiny tangential force on the electron is $F_e = F v_p / 2c = F r_p \Omega / 2c$.

The magnitude of the tiny tangential force on the proton is $F_p = F v_e / 2c = F r_e \Omega / 2c$.

The magnitudes of the torques on the electron and on the proton are $T_e = r_e F_e$ and $T_p = r_p F_p$, both equal to $F r_e r_p \Omega / 2c$.

The total torque on the electron-proton system is $T_{\text{total}} = T_e + T_p = 2T = F r_e r_p \Omega^2 / c$.

The torque power delivered to the system is $P_{\text{torque}} = T\Omega = F r_e r_p \Omega^2 / c$.

The squared orbit frequency is determined from either $F = m_e r_e \Omega^2$ or $F = m_p r_p \Omega^2$.

The more convenient of the two options is $\Omega^2 = F / m_e r_e$. With that expression, the approximation $r_e \approx r_e + r_p$ yields:

$$P_{\text{torque}} = Fr_{\text{e}}r_{\text{p}}\Omega^2/c = F^2r_{\text{p}}/m_{\text{e}}c = \left(e^4r_{\text{p}}/m_{\text{e}}c\right)/\left(r_{\text{e}}+r_{\text{p}}\right)^4 \approx \left(e^4/m_{\text{p}}c\right)/\left(r_{\text{e}}+r_{\text{p}}\right)^3 . \tag{12}$$

This is a reasonably simple expression. But more important than its simplicity is its very *existence*. The existence of any such expression means that there exists an energy *gain* mechanism to balance against the known energy loss mechanism, *i.e.* radiation. This situation provides a chance for balance, allowing the Hydrogen atom to avoid death by energy loss to radiation.

The P_{torque} is actually quite large, and so it changes the whole emphasis of worry concerning the Hydrogen atom. The question becomes, not why does the Hydrogen atom not radiate and collapse to death, but rather why does the Hydrogen atom not torque itself up and expand way beyond its known size?

The fact is: there exists much, much more radiation than was previously worried about, and it is enough to produce the proper balance between radiation and torque.

In [17], I said that the extra radiation arises from finite signal propagation speed, which results in circular motion of the center of mass of the Hydrogen atom, which in turn produces Thomas rotation, and thereby scales up by a factor of 2 the overall rotation rate generating the radiation, which increases the radiation power by a factor of 2^4.

But what should one say about that center-of-mass circular motion? In Newtonian physics, where signal propagation speed is infinite, there is no such thing. In Maxwell physics, the emphasis is on fields, and the responses of individual charges, but not as much on the responses of whole charge systems, such as atoms. So the issue doesn't come up there. In Einstein's relativity physics, the emphasis is often on the observers of events more than on the events themselves. System center-of-mass circulation seems not to come up, although Thomas rotation does.

In [17], I noted that Thomas rotation is generally believed to be a result of the properties of Lorentz transformations, and hence of SRT. That is the belief because one can think of Lorentz transformations, not only in the usual, passive sense, as conversion from an observer in one inertial coordinate frame to another observer in another inertial coordinate frame, but also in the active sense, as the application of a 'boost' in velocity, the result of an acceleration, the result of a physical force. A series of non-co-linear boosts does indeed produce Thomas rotation.

But in [17] I also remarked that Thomas rotation does arise, not just from Lorentz transformations, but also from Galilean Transformations. That fact can be demonstrated in detail as follows:

For simplicity, let all motion be in the x, y plane. The scenario begins at time coordinate ct_0 with one of the particles, say the electron, at rest at spatial coordinates x_0, y_0. Let an attraction from another particle act in the x direction. Let an increment of velocity $\Delta\mathbf{V}_x = \Delta V$ be imposed, and let an increment of time Δt elapse. The coordinates of the electron then become:

$$t_1 = c(t_0 + \Delta t), x_1 = x_0 + \Delta V \Delta t \text{ and } y_1 = y_0. \tag{13}$$

Now let an attraction from another particle act in the y direction. Let an increment of velocity $\Delta \mathbf{V}_y = \Delta V$ be imposed, and let another increment of time Δt elapse. The coordinates of the electron then become:

$$\begin{aligned} &t_2 = c(t_0 + \Delta t + \Delta t), \\ &x_2 = x_0 + \Delta V(\Delta t + \Delta t) = x_0 + 2\Delta V \Delta t, \\ &\text{and } y_2 = y_0 + \Delta V \Delta t. \end{aligned} \tag{14}$$

Observe that, if the Galilean velocity boosts had been applied in the opposite order, then the ending coordinates of the electron would have been:

$$\begin{aligned} &t_2 = c(t_0 + \Delta t + \Delta t), \\ &x_2 = x_0 + \Delta V \Delta t, \\ &\textit{and } y_2 = y_0 + 2\Delta V \Delta t. \end{aligned} \tag{15}$$

Observe that: the squared incremental length changes have the same magnitude either way: $(2\Delta V\Delta t)^2 + (\Delta V\Delta t)^2 \equiv (\Delta V\Delta t)^2 + (2\Delta V\Delta t)^2 = 5(\Delta V\Delta t)^2$. That fact means the two possible sequences of Galilean boost applications differ only by a rotation. That means each one individually contains a rotation equal to half that total angle difference. This is the Thomas rotation.

Let me now go further, and assert that Thomas rotation will arise from *any* kind of velocity transformation – Lorentz, or Galilean, or any other new kind that may not have a name yet. Thomas rotation is a property of actual reality, not of any particular mathematical model for reality.

With the Thomas rotation included, the total radiation from the atomic system is:

$$P_{\text{total radiated}} = \underline{2^4} \frac{2e^2}{3c^3} a_e^2 = \left(\underline{2^5} e^6 / m_e^2\right) / 3c^3 (r_e + r_p)^4. \tag{16}$$

The value of the separation $r_e + r_p$ for which $P_{\text{total radiated}} = P_{\text{torque}}$ is:

$$r_e + r_p = 32 m_p e^2 / 3 m_e^2 c^2 = 5.5 \times 10^{-9} \text{cm}. \tag{17}$$

In the traditional approach to QM, $r_e + r_p = h^2 / 4\pi^2 \mu e^2$, where μ is the reduced mass, defined by $\mu^{-1} = m_e^{-1} + m_p^{-1}$, and very nearly equal to m_e, and h is Planck's constant, 6.626176×10^{-34} Joule-sec, a fundamental constant given by Nature.

The present analysis does not require Planck's constant as an input. Instead, it provides an estimate of Planck's constant as an *output*:

$$h = \sqrt{4\pi^2 \mu e^2 (r_\mathrm{e} + r_\mathrm{p})} \approx \sqrt{4\pi^2 m_\mathrm{e} e^2 32 m_\mathrm{p} e^2 / 3 m_\mathrm{e}^2 c^2}$$
$$\approx \frac{\pi e^2}{c} \sqrt{128 m_\mathrm{p} / 3 m_\mathrm{e}} \approx 6.77 \times 10^{-34} \,\text{Joule-sec} \quad .; \tag{18}$$

and, for convenience, also an estimate of the often-seen reduced Planck's constant:

$$\hbar = h / 2\pi \approx (e^2 / c)\sqrt{32 m_\mathrm{p} / 3 m_\mathrm{e}} \approx 1.08 \times 10^{34} \,\text{Joule-sec.} \tag{19}$$

These estimates can be improved by taking due account of the fact that the sines of small angles are not exactly equal to the angles themselves, and the cosines of small angles are not exactly equal to unity. The sine corrections are third order in angle, while the cosine corrections are second order in angle, which is more significant. Including those corrections reduces the radiation power slightly, and so reduces the solution $r_\mathrm{e} + r_\mathrm{p}$ slightly, and so reduces the estimates of h and $\hbar$ slightly – a step in the right direction.

Observe that, in making h an output *from*, rather than an input *to*, a theory, the present work follows both Enders [18] and Ralston [19].

5. EM interactions between neutral atoms: A candidate model for gravity?

Our current best understanding of gravity comes from GRT, which is founded in, and developed from, SRT. The parameter *c* from SRT appears in GRT, and reveals the lineage. Like electromagnetic signals, gravitational signals must have the finite propagation speed *c*. So might gravitational signals actually be electromagnetic signals? If we choose to update SRT with a more realistic signal model in place of the number *c*, does that require a similar update for GRT? This Section explores such questions.

First let it be noted that the Einstein gravitational field equations, like Maxwell's coupled field equations, are a description at the microscopic scale, but the observable phenomena exist at an extremely macroscopic scale.

Figure 2 illustrates a fairly typical barred spiral galaxy. What this image seems to suggest is: there are mass concentrations at the two armpits of this galaxy. Maybe they are mega stars, or black holes. Maybe they orbit another mass concentration at the center of the galaxy, or maybe they just orbit each other. In any event, the two mass concentrations together create a rather structured field of gravitational potential, into which millions, or billions, of smaller stars are entrained, or temporarily detained, as they orbit the galaxy.

Figure 3 shows the skeleton of a potential field created by two super-massive bodies orbiting at half the signal propagation speed. The lines mark minima in gravitational potential as a function of angle around the galaxy. Observe that this skeleton approximately matches this galaxy image.

Figure 2. A typical barred spiral galaxy: "A Barred Spiral Galaxy ngc 1300 hubble photo".

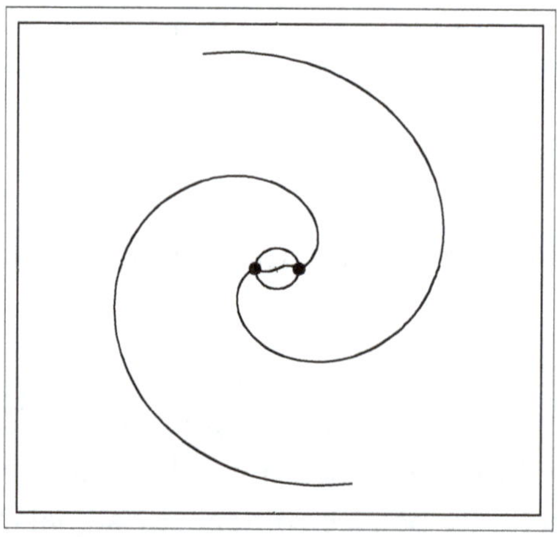

Figure 3. The skeleton for a barred spiral disc galaxy. Originally computed for [20].

Note 1: To persist over time, the orbit speed of the driving two-body system in Figs. 2 and 3 must be such that the rates of energy loss by gravitational radiation and energy gain by torquing balance each other. Like the Hydrogen atom, the galactic-size two-body problem can be solved this way.

Note 2: Since most of the captive stars in Fig 2 orbit at lesser speed than the driving two bodies, and their potential pattern, with its skeleton, Fig. 3, the individual stars in the outer reaches do not keep up with the rotating spiral potential pattern. An individual star sees a recurring 'density wave' of neighbor stars first approaching, then receding. Such density waves have long been known, but not well explained.

Note 3: Fig. 3 is constructed using brute-force calculation of potential for lots and lots of location points, and then using numerical search over angles for local minimum values, and then fitting a function to the minima. Note the slightly sinuous bar between the two driving bodies. Even this detail is suggested in Fig. 2.

Now, in order to model gravity in terms of electromagnetic interactions, we need an expression of electromagnetic interaction that is appropriate for neutral atoms. The best expression appears to be one that was known even *before* Maxwell. André Marie Ampère already had a well-developed theory about forces between what he called 'current elements'. This term referred to charge-neutral material increments in electrical circuits. In modern times, P. Graneau wrote extensively about Ampère's theory and experiments; see for example Graneau [21]

Ampère's theory works perfectly well for ordinary closed circuits, as well as for incomplete broken circuits, such as may exist momentarily in transient situations, like explosive rupture of circuits. Ampère's theory ought not be forgotten solely on the basis that more modern theory also works perfectly well for closed and stable electrical circuits. Indeed, in some technological applications involving transient situations like ruptures, Ampère's theory explains more than the modern theory does.

One reason why Ampère's formulation can sometimes be more powerful than a formulation based on Maxwell's equations is that Ampère's formulation can describe a multi-participant scenario straight away, whereas Maxwell's equations require iteration through successive steps involving both Maxwell's equations, and the Lorentz force law for the force $\mathbf{F}$ acting on a charge q moving with velocity $\mathbf{v}$:

$$\mathbf{F} = q\left[\mathbf{E} + \mathbf{v} \times \mathbf{B}\right]. \tag{20}$$

The iteration goes as follows: first, the input charge positions and motions generate $\mathbf{E}$ and $\mathbf{B}$ fields; second, the Lorentz force law tells how each charge q responds to the fields from the other charges; another step through Maxwell's equations tells how all the fields change, and so on.

One particular scenario illustrates the difference between the approaches especially well. Consider a current-carrying wire. It is tedious to use the Maxwell-Lorentz-Maxwell-Lorentz iterative approach to arrive at the understanding that the moving electrons favor the surface of the wire, or even the exterior neighborhood near the wire, leaving the interior of the wire depleted of electrons, and therefore in a state of internal repulsion between the remaining positive nuclei. The Ampère's formulation skips over all this detail, and describes the resulting consequence: the wire experiences internal longitudinal force, and in fact might even rupture. If it does rupture, one can easily tell that the event was *not* due to ordinary resistive heating and melting, since the fragments are found to be neither hot to touch nor melted in appearance.

The Ampère approach looks promising for gravity problems because *any* gravity problem is definitely a multi-participant scenario. And for the same reason, the following analysis also invokes ideas from modern Statistical Mechanics.

Ampère's force formula can be written:

$$\Delta F_{m,n} = +i_m i_n \left[\Delta m \, \Delta n / (r_{m,n})^2 \right] (3\cos\alpha \cos\beta - 2\cos\gamma). \tag{21}$$

The indices m and n identify two interacting currents. The i_m and i_n are current magnitudes. The Δm and Δn are magnitudes of tiny directed length increments $\Delta \mathbf{m}$ and $\Delta \mathbf{n}$ through which the currents flow. The products of currents and directed length increments, $i_m \Delta \mathbf{m}$ and $i_n \Delta \mathbf{m}$, are the current elements. The $r_{m,n}$ is the length of the vector separation $\mathbf{r}_{m,n}$ between the current elements. The α, β, and γ are angles with respect to the connecting line between the two current elements, and with respect to each other. Current element $i_m \Delta \mathbf{m}$ is at angle α from the connecting line, and current element $i_n \Delta \mathbf{n}$ is at angle β from the connecting line. The γ is the angle between the two planes defined by the connecting line and each of the two current elements, as if the distance $r_{m,n}$ did not separate them. The value ranges are all full circle: $0<\alpha<2\pi$,

One can get a feel for the general behavior of Ampère's force formula by considering the angle factor $3\cos\alpha\cos\beta - 2\cos\gamma$ for a few special cases:

1. Current elements side-by-side and parallel, as in parallel wires. Both current elements are perpendicular to the connecting line, so α and β are $\pi/2$ and $\cos\alpha$ and $\cos\beta$ are zero. But γ is zero, and $\cos\gamma=1$, so the angle factor evaluates to -2. The force $\Delta F_{m,n}$ is then negative. The current elements attract each other. If they reside in parallel wires, the wires attract each other. This you know from experience is true. In a plasma, instead of in a solid wire, it is called the 'pinch effect'.

2. Current elements side-by-side, but anti-parallel. This case is just opposite to Case 1 above: now $\gamma=\pi$ and $\cos\gamma=-1$. The current elements repel each other. If they reside in a circuit, that circuit likes to straighten out any kinks and enclose more area. This you may know from experience is true.

3. Current elements end-to-end, as in an electrical circuit. All three angles are zero, all three cosines are unity, and the angle factor evaluates to +1, so the force $\Delta F_{m,n}$ is positive. The current elements repel each other.

The $1/(r_{m,n})^2$ aspect of the Ampère force law is just like Newton's law for gravity. Ampère designed his law that way, because, in his time, the greatest prior achievement in Science was Newton's conquest of gravity. Now we wish to return the favor, and exploit the Ampère Force Law to understand something novel about gravity.

What makes the Ampère current element so potentially appropriate for application to gravity? First of all, it is charge-neutral, like the masses in a gravity scenario. Secondly, its electrons are moving, and although its nucleus is moving too, that motion is not anywhere near as fast. So at all times and all places where matter exists, at the microscopic level a net electron current flows.

The concept that current elements generate forces that can attract or repel each other suggest that pairs of current elements – or pairs of atoms – can be regarded as a system that can have positive or negative total energy. The kinetic part of the energy may be disregarded, since the current elements may be essentially static, but the potential part of the energy is worth paying attention to.

The main novel feature that gravity presents is that we usually have, not two current elements, but huge numbers of atoms, and each atom must have some relationship with *all* other atoms. The complexity of the situation naturally conjures up ideas from Statistical Mechanics. Here, ideas from Statistical Mechanics are applied to gravity described in terms of Ampère forces between atoms that are viewed as current elements. Some atom-to-atom relationships are momentarily attractive, and some relationships are momentarily repulsive, and all relationships must vary over time. We can look at the population of atom pairs as a whole, and think of it as a statistical ensemble, in which every condition of attraction/repulsion is represented somewhere.

Every area of physics that has statistical ensembles has Gaussian probability functions. In Classical Thermodynamics, a Gaussian probability density function for a random variable, such as a component of a particle momentum vector, implies maximum entropy, subject to a prescribed value for the standard deviation of that random variable. In Quantum Mechanics, a Gaussian probability density function (the squared wave function amplitude) is associated with minimum uncertainty, meaning minimum product of standard deviations in Fourier conjugate variables, like position and momentum.

Sometimes it is not immediately obvious that a problem involves the equivalent of a Gaussian function, because the Gaussian itself involves a squared variable, such as x^2 or p^2. The squared variable is proportional to some energy E. So one sees probability density functions expressed in the form $\exp(-E/<E>)/<E>$ where $<E>$ is the average value of energy E, usually something like kT, where k is Boltzmann's consistent and T is absolute temperature.

In the case of gravity, that energy E of interest is gravitational potential energy. It can have both positive *and* negative values. The central concept in Statistical Mechanics is that lower-energy states are populated more richly than higher-energy states are. This concept means that any two atoms, viewed as current elements, will be with respect to each other in a state of negative potential energy more often than in a state of positive potential energy. So they will, on average, attract each other more than repel each other. Therefore, $<E>$ is negative. Since temperature cannot be negative, this is something novel.

To deal with negative $<E>$ we really need a probability density function with a Gaussian factor of the form $\exp[-E^2/2<E^2>]$, where $<E^2>$ is another parameter, positive, but also not related to temperature.

A few examples can illustrate how to find the parameters.

Let the two energies be E_{max} and $E_{min} = -|E_{max}|$. With the attractive–force, negative-energy state dominating the scenario, $<E>$ is negative, $<E> = -|<E>|$. The two Boltzmann factors are:

$$\exp(+E_{max}/|<E>|)\text{for the state with negative energy, } E_{min} = -E_{max} \tag{22}$$

and

$$\exp(-E_{max}/|<E>|)\text{for the state with positive energy,} E_{max}. \tag{23}$$

The average energy E_{avg} must satisfy the definition:

$$E_{avg} = \frac{-E_{max}\exp(+E_{max}/|<E>|) + E_{max}\exp(-E_{max}/|<E>|)}{\exp(+E_{max}/|<E>|) + \exp(-E_{max}/|<E>|)}. \tag{24}$$

Simple trial calculations and numerical search of the results works well enough to solve the problem at hand. The solution is approximately:

$$E_{avg} \approx -E_{max}/1.2 = -0.8333E_{max} = -\tfrac{5}{6}E_{max}. \tag{25}$$

(There is also, of course, a positive, and presently irrelevant, solution of the same magnitude.)

For this case we have we have line integration in place of point evaluation. E_{avg} becomes:

$$E_{avg} = -\int_{-E_{max}}^{0} E\sinh(E/|<E>|)dE \Big/ \int_{-E_{max}}^{0} \cosh(E/|<E>|)dE. \tag{26}$$

Because the denominator is simpler, begin with that. It is:

$$\begin{aligned} |E_{avg}|\sinh(E/|E_{avg}|)\Big|_{-E_{max}}^{0} &= -\big[|E_{avg}|\sinh(-E_{max}/|E_{avg}|)\big] \\ &= +|E_{avg}|\sinh(E_{max}/|E_{avg}|) \;. \end{aligned} \tag{27}$$

This is a positive number.

The numerator is more complicated, but it can be evaluated using integration by parts:

$$\int_{-E_{max}}^{0} UdV = UV\Big|_{-E_{max}}^{0} - \int_{-E_{max}}^{0} VdU \;. \tag{28}$$

where $U = E$ and $dV = \sinh(E/|E_{avg}|)dE$, so that $V = |E_{avg}|\cosh(E/|E_{avg}|)$. The first term in the numerator evaluation is:

$$UV\big|_{-E_{\max}}^{0} = -U(-E_{\max})V(-E_{\max}) = -\left[-E_{\max}\,|<E>|\cosh(-E_{\max}/|<E>|)\right] \\ = E_{\max}\,|<E>|\cosh(E_{\max}/|<E>|) \quad . \tag{29}$$

The second term in the numerator evaluation is:

$$-\int_{-E_{\max}}^{0} VdU = -\int_{-E_{\max}}^{0} |<E>|\cosh(E/|<E>|)dE = -|E_{avg}|^2 \sinh(E/|<E>|)\big|_{-E_{\max}}^{0} \\ = |<E>|^2 \sinh(-E_{\max}/|<E>|) = -|<E>|^2 \sinh(E_{\max}/|<E>|) \quad . \tag{30}$$

.

The sought numerator divided by denominator for $<E>$ is then:

$$<E> = \frac{E_{\max}\,|<E>|\cosh(E_{\max}/|<E>|) - |E_{avg}|^2 \sinh(E_{\max}/|<E>|)}{|E_{avg}|\sinh(E_{\max}/|E_{avg}|)} \\ = \frac{E_{\max}}{\tanh(E_{\max}/|<E>|)} + |<E>| \quad . \tag{31}$$

Since the sought $<E>$ is negative, equal to $-\,|<E>|$, we have:

$$2<E> = -2|<E>| = \frac{-E_{\max}}{\tanh(E_{\max}/|<E>|)}. \tag{32}$$

Again, numerical search is a practical approach for finding a solution. We find:

$$<E> \approx -0.485E_{\max}. \tag{33}$$

Observe that, as should be expected, this solution is significantly smaller in magnitude than was the solution with only two energy values, $\pm E_{\max}$, to balance between, which came in at $-0.833E_{\max}$.

The continuous Gaussian profile,

$$\exp(-E^2/2<E^2>)\big/\sqrt{2\pi<E^2>}, \tag{34}$$

extends to infinity in both directions. This attribute is inappropriate for the problem at hand, which definitely possesses limits $\pm E_{\max}$ beyond which the modeling problem does not extend.

Therefore, let us turn to discrete approximations for a Gaussian. These are based on the binomial expansion for an arbitrary $(a+b)^n$. The binomial coefficients are familiar to many people from Pascal's famous triangle:

$$\begin{array}{c} 1 \\ 1 \quad 1 \\ 1 \quad 2 \quad 1 \\ 1 \quad 3 \quad 3 \quad 1 \\ etc. \end{array} \tag{35}$$

The numbers in Pascal's triangle are constructed with addition of neighboring numbers above. This is easy for small n, but small n means *crude*, and we need *refined*. So we want *large* n. So we need a formula involving multiplication instead of addition. That would be:

$$1,\ n,\ \frac{n(n-1)}{1\times 2},\ \frac{n(n-1)(n-2)}{3!},\ \cdots,\ \frac{(n)!}{(n')!(n-n')!},\ \cdots,\ \frac{n(n-1)(n-2)}{3!},\ \frac{n(n-1)}{1\times 2},n,1. \tag{36}$$

Observe that the binomial coefficients are symmetric around the middle of the list, like a Gaussian function is symmetric around zero argument. If n is an even number, the number of binomial coefficients is $2n-1$, odd, and the middle one, the maximum one, is $n!/[(n/2)!]^2$. If n is an odd number, the number of binomial coefficients is $2n$, an even number, and the middle two numbers, the maximum two, are both $n!/\{[(n+1)/2]![(n-1)/2]!\}$.

For our modeling problem, let n be an odd number. Let the binomial coefficients be represented as B_b with $b=-n$ to $b=1$. Let them be associated with equal energy increments $\Delta E = E_{max}/n$ starting from $-E_{max}$ and covering the range to zero. Associate the minimum binomial coefficient B_{-n} with the increment starting with $-E_{max}$, and the maximal binomial coefficient with the increment ending with $E=0$, and associate the other coefficients with the increments between those limits. The problem to solve is:

$$<E>\approx \sum_{b=-n}^{1} B_b E_b \sinh(E_b/<E>) \Big/ \sum_{b=-n}^{1} B_b \cosh(E_b/<E>). \tag{37}$$

Numerical investigations done to date suggest that the solution comes at approximately $<E>=-E_{max}/2\sqrt{n}$. Here the $\sqrt{n}$ for this discrete model is analogous to the standard deviation σ for the corresponding continuous Gaussian.

The problem of modeling gravity therefore reduces to the problem of determining what value of n should be used. Here is the most pertinent fact: compared to anything electromagnetic, gravity is extremely weak. Consider two Hydrogen atoms at a given separation distance. Let us compare the gravitation force with the maximum Ampère force between them.

The gravitational attraction is proportional to $G(m_p)^2$, where G is the universal gravitation constant, about 6.6×10^{-11} Newton× meter^2 per kilogram^2, and m_p is the mass of the proton, about 1.66×10^{-27}kg, so $(m_p)^2$ is about 2.76×10^{-54} kg^2. Overall,

$$\frac{1}{4\pi\varepsilon_0}e^2(v/c)^2 \approx \frac{2.56\times10^{-38}\times0.45\times10^{-4}}{1.113\times10^{-10}} \approx 1.035\times10^{-32}\,\text{Newton}\times\text{meter}^2. \tag{38}$$

Clearly, the maximum Ampère force between atoms viewed as current elements is generously larger than the gravitational force between atoms viewed as charge-neutral masses – by about 32 orders of magnitude!

But the typical Ampère force is nowhere near as big as the maximum Ampère force, due to the fact that it depends on three angles, any one of which can spoil it. Occurrence of the maximum Ampère force is very rare indeed. And occurrence of the *minimum* (negative) Ampère force is equally rare. Only the piddling near-zero Ampère forces are common, and even then, each tiny attractive force is mostly cancelled with the tiny repulsive force of equal magnitude but less frequent occurrence. All we have is a tiny residue of attractive force, due to the nature of Boltzmann factors and Statistical Mechanics.

Of course, being tiny does not mean being insignificant. Like the tiny residue that is microwave background radiation, the tiny residue that is gravity is a possible key to understanding something about the Universe in which we live. Here is an example problem: at present, we know the actual particle radius of the electron is something extremely tiny, but we do *not* know what its numerical value is. There exist a number of length-dimensioned quantities associated with the electron, all called 'radius', but distinguished by specific names and numerical values. MacGregor [22] lists seven of them. Most are on the order of 10^{-13} cm, although one is much smaller, and is presently only upper-bounded at $<10^{-16}$ cm.

The radius attributed to the electron can have a role in the gravity problem. The ratio of an atomic radius to the electron radius can imply a candidate level of discretization for the binomial approximation to the Gaussian factor involved in the gravity problem. For an atomic radius, let us consider the first orbit radius of Hydrogen, $r_{H1}=0.529\times10^{-8}$cm. For the electron radius, let us consider two of the possibilities from MacGregor [22].

One of the electron radii is called *classical*. This one captures the Coulomb energy equivalence

$$\frac{1}{4\pi\varepsilon_0}e^2/r_{\text{e classical}} = m_e c^2, \tag{39}$$

which implies

$$r_{\text{e classical}} = \frac{1}{4\pi\varepsilon_0}\left(e^2/m_e c^2\right) = 2.82\times10^{-13}\,\text{cm}. \tag{40}$$

The ratio $\rho = r_{H1} / r_{e\ \text{classical}}$ is then:

$$n = 0.529\times10^{-8}/2.82\times10^{-13} = 5.29\times10^{-9}/2.82\times10^{-13} \approx 1.86\times10^{4}, \tag{41}$$

which implies

$$n = 0.529\times10^{-8}/2.82\times10^{-13} = 5.29\times10^{-9}/2.82\times10^{-13} \approx 1.86\times10^{4}. \tag{42}$$

The square root of this number would then be the dimensionless $\sqrt{n}$ for the discretization:

$$\sqrt{n} = \sqrt{1.86\times10^{4}} \approx 1.37\times10^{2} = 137. \tag{43}$$

This is a number already famous in Physics, but in a context other than gravity. It is the inverse of the so-called 'fine structure constant' α, defined as:

$$\alpha = 2\pi e^{2} / ch. \tag{44}$$

This number plays a role in spectroscopy, where spectral lines occur in families, closely spaced but clearly distinguishable. There, the explanation comes from QM; clearly, another manifestation of natural discretization.

But if the classical radius of the electron were used in the gravity problem, the ratio of the average Ampère force magnitude to the maximum Ampère force magnitude would be approximately

$$\sqrt{n}/2n = 1/2\sqrt{n} \approx 3.65\times10^{-3}. \tag{45}$$

This ratio is *not* appropriately small, so this is *not* the right discretization level for the gravity problem.

Another one of the electron radii given by MacGregor [22] is called *actual*. It characterizes results of scattering experiments, and is the one presently only upper-bounded, at $r_{e\ \text{actual}} < 10^{-16}$ cm. No one knows how much smaller it could eventually turn out to be. So how much smaller would it *have* to be, in order to account for the extreme weakness of gravity? Gravity requires $2/\sqrt{n} \approx 10^{-32}$, or.

$$\sqrt{n} \approx 2\times10^{+32}, \text{ or } n \approx 4\times10^{64}. \tag{46}$$

That in turn requires:

$$r_{\text{e actual}} = r_{\text{H1}} / \rho = 0.529 \times 10^{-8}\text{cm} / 4 \times 10^{64} \approx 10^{-73}\text{cm}. \tag{47}$$

At present, such a value for $r_{\text{e actual}}$ certainly looks impossible to test with any kind of measurement. It is smaller than anything we yet know about any elementary particle. But that circumstance may be a good thing, because an extremely small electron makes it easier to understand what data from Chemistry reflects, discussed next. And an extremely small electron, along with a correspondingly small positron, helps explain aspects Elementary Particle Physics, discussed after that.

6. Algebraic chemistry and EZ water

Prof. Gerald Pollack of U. Washington wrote the authoritative book [23] about the physical phenomenon called 'EZ water'. The EZ is short for Exclusion Zone, with the word 'exclusion' referring to a surface phenomenon that expels positive hydronium ions.

Prof. Pollack gave a talk about EZ Water at the 2013 meeting of the Natural Philosophy Alliance at College Park, MD, USA. All the phenomena he described were surprising; some were truly puzzling. EZ water apparently makes extended orderly arrays of hexagonal units. How can *that* behavior comport with our understanding that Nature maximizes entropy? Explanations then available were not at all quantitative. That fact suggested a real need for a more quantitative approach.

I had recently written my book about Algebraic Chemistry (AC). [24] The name reflects the fact that the technique has no integrals or other complicated math operations that would demand capabilities beyond those of a hand calculator. The worst operation is square root. So the AC approach looked promising for quick application to EZ water.

The fundamental idea behind AC is that all atoms share some similarities with Hydrogen atoms: **1)** They have a nucleus that is similar to a proton, but scaled up to nuclear charge Z and nuclear mass M ; **2)** They have a population of electrons that is not entirely unlike a single electron; *i.e.*, an interacting community that is somewhat coherent, and somewhat like one *big* electron orbiting the nucleus; **3)** It is possible for the electron count to be different from the nuclear charge. This last possibility is what characterizes ions, and thereby creates all of Chemistry.

We begin with a clue: Eq. (17) indicates that the radius of the Hydrogen atom scales with the mass of the proton. This fact suggests that the base orbit energy of the Hydrogen atom scales with the inverse of proton mass. It further suggests that for element with nuclear charge Z and mass M, the base orbit energy may scale with Z/M. If so, then when first-order ionization potentials for all elements are scaled by the inverse factor, M/Z, then the scaled first-order ionization potentials (called $IP_{1,Z}$) might fall into some pattern.

We proceed with an observation: A pattern indeed emerges: the rise on *every* period in the Periodic Table is *exactly the same factor*, $7/2$.

We make a Hypothesis: All $IP_{1,Z}$ contain information valuable for all other elements: *population generic information*. Each $IP_{1,Z}$ contains a universal baseline contribution $IP_{1,1}$ about interaction between the nucleus and the population of electrons as a whole. For all elements beyond Hydrogen, there is also a contribution $\Delta IP_{1,Z}$ about interactions among the electrons.
The $\Delta IP_{1,2}$ can be very significant. For Helium, $\Delta IP_{1,2}$ is *huge*, meaning that two electrons bond together very *strongly*. And for Lithium, $\Delta IP_{1,3}$ is *negative*, meaning that two electrons actively work together to try to exclude a third electron.

Over the periods, there is obvious detail about the electron-electron interactions. Within each period, there are obvious sub-periods keyed to the nominal angular-momentum quantum number that is being filled. Plotted on a log scale, all sub-period rises are straight lines. The slopes all appear to be rational fractions. We can display these rational fractions in a Table, as was done in [24] and [25]:

period	N	l	*fraction*	l	*fraction*	l	*fraction*	l	*fraction*
1	1	0	1						
2	2	0	1/2	1	3/4				
3	2	0	1/2	1	3/4				
4	3	0	1/4	2	5/18	1	2/3		
5	3	0	1/4	2	5/18	1	2/3		
6	4	0	1/4	3	7/48	2	5/16	1	9/16
7	4	0	1/4	3	7/48	2	5/16	1	9/16

A non-traditional parameter N is included in the display because, for $l>0$, it is possible to write a simple formula for the fraction:

$$\text{fraction} = \left[(2l+1)/N^2\right]\left[(N-l)/l\right]. \tag{48}$$

Also, all periods in the Periodic Table have length $2N^2$.

All this numerical regularity suggests that there really is a reliable pattern here, and we can reasonably seek to exploit it. Here is the first exploitation that suggests itself: Given first-order ionization potentials of many elements, we can estimate the additional energy required to remove a second electron from each, and then a third, and so on. This was first done in [24]. Formulae were given for each *individual* electron removal or addition, and evaluated for a large number of elements.

One point that Ref. [24] emphasized was that the energy to remove a second electron, or a third, and so on, is *not* the same thing as the so- called 'second-order ionization potential', 'third-order ionization potential', and so on. Those energies are very *large*, which implies that those events are very *violent*: ripping two, or three, or more, electrons off an atom *all at once*.

Those energies do exhibit a lot of numerical regularity, but that isn't important for understanding typical lab-bench chemistry, which is all about *gentle* events that occur *one-at-a-time.*

Ref [25] presented the equivalent summed formulae for removal of, or addition of, one, two, and three electrons, removed or added *one-at-a-time*. Basically, use of these formulae save the user some repetitive arithmetic that would be incurred using the formulae from Ref. [24].

Development of More Formulae:

For the present paper, the formulae from [25] are extended from the illustrative cases of removing, or adding, one, two, and three electrons, to the general case of removing, or adding, N electrons.

Ref. [25] used symbols W and H to distinguish between energy increments associated with electron-nucleus interaction, and energy increments associated with electron-electron interactions. That distinction is analogous to the distinction between work and heat in thermodynamics: the work part is something a human can control, and the heat part is something that Nature simply does, regardless of what the human does.

We had:

$$W_{\text{removing } e_1 \text{ from the neutral atom}} = IP_{1,1}(Z / M_Z). \tag{49}$$

The ion produced has a little less attraction between the nucleus and the now reduced electron cloud. So removing another electron should take a little less work:

$$W_{\text{removing } e_1 \ \& \ e_2} = IP_{1,1}\left[Z + \sqrt{Z(Z-1)}\right] / M_Z. \tag{50}$$

And then:

$$W_{\text{removing } e_1,\, e_2,\, \& \ e_3} = IP_{1,1}\left[Z + \sqrt{Z(Z-1)} + \sqrt{Z(Z-2)}\right] / M_Z. \tag{51}$$

This pattern generalizes to:

$$W_{\text{removing } e_1 \text{ through } e_N} = IP_{1,1}\left[\sum_{i=1}^{N} \sqrt{Z(Z+1-i)}\right] / M_Z. \tag{52}$$

We also had:

$$H_{\text{removing } e_1 \text{ from the neutral atom}} = (\Delta IP_{1,Z} - \Delta IP_{1,Z-1})(Z / M_Z). \tag{53}$$

We inferred in [25] that:

$$H_{\text{removing } e_1, \& e_2} = \left[\Delta IP_{1,Z} Z - \Delta IP_{1,Z-2}(Z-2) \right] / M_Z. \tag{54}$$

and

$$H_{\text{removing } e_1, e_2, \& e_3} = \left[\Delta IP_{1,Z} Z - \Delta IP_{1,Z-3}(Z-3) \right] / M_Z. \tag{55}$$

This pattern generalizes to:

$$H_{\text{removing } e_1 \text{ through } e_N} = \left[\Delta IP_{1,Z} Z - \Delta IP_{1,Z-N}(Z-N) \right] / M_Z. \tag{56}$$

Finally, in [25] we had:

$$\begin{gathered}(W+H)_{\text{removing } e_1, \& e_2} = \\ IP_{1,1}\left[Z + \sqrt{Z(Z-1)} \right] / M_Z + \left[\Delta IP_{1,Z} Z - \Delta IP_{1,Z-2}(Z-2) \right] / M_Z \quad ,\end{gathered} \tag{57}$$

and

$$\begin{gathered}(W+H)_{\text{removing } e_1, e_2, \& e_3} = \\ IP_{1,1}\left[Z + \sqrt{Z(Z-1)} + \sqrt{Z(Z-2)} \right] / M_Z + \left[\Delta IP_{1,Z} Z - \Delta IP_{1,Z-3}(Z-3) \right] / M_Z \quad ,\end{gathered} \tag{58}$$

This pattern generalizes to:

$$\begin{gathered}(W+H)_{\text{removing } e_1 \text{ through } e_N} = \\ IP_{1,1}\left[\sum_{i=1}^{N} \sqrt{Z(Z+1-i)} \right] / M_Z + \left[\Delta IP_{1,Z} Z - \Delta IP_{1,Z-N}(Z-N) \right] / M_Z \quad ,\end{gathered} \tag{59}$$

Now let us turn to *adding* electrons. First, use the formula for the energy for removing an electron from a neutral atom of element Z to describe instead *removing* an electron from the *singly charged negative ion* of element Z, which has $Z+1$ electrons to start with:

$$(W+H)_{\text{removing } e_1 \text{ from negative ion}} = \left[IP_{1,1}\sqrt{Z(Z+1)} + \Delta IP_{1,Z+1}(Z+1) - \Delta IP_{1,Z} Z \right] / M_Z. \tag{60}$$

Reversing the direction of the operation:

$$(W+H)_{\text{adding } e_1 \text{ to neutral atom}} = -\left[IP_{1,1}\sqrt{Z(Z+1)} + \Delta IP_{1,Z+1}(Z+1) - \Delta IP_{1,Z}Z \right] / M_Z . \tag{61}$$

This means the work for adding one electron into the nuclear field is:

$$W_{\text{adding } e_1 \text{ to neutral atom}} = -IP_{1,1}\sqrt{Z(Z+1)} / M_Z . \tag{62}$$

And the heat for re-adjusting the electron population is:

$$H_{\text{adding } e_1 \text{ to neutral atom}} = -\left[\Delta IP_{1,Z+1}(Z+1) + \Delta IP_{1,Z}Z \right] / M_Z . \tag{63}$$

Now let us add a second electron. This will require additional work:

$$W_{\text{adding } e_2 \text{ after } e_1} = -IP_{1,1}\sqrt{Z(Z+2)} / M_Z . \tag{64}$$

And it will cause another heat adjustment:

$$H_{\text{adding } e_2 \text{ after } e_1} = -\Delta IP_{1,Z+2}(Z+2) / M_Z + \Delta IP_{1,Z+1}(Z+1) / M_Z . \tag{65}$$

This means total energy involved in adding two electrons is:

$$\begin{aligned} &(W+H)_{\text{adding } e_1 \ \& \ e_2} = \\ &-IP_{1,1}\left[\sqrt{Z(Z+1)} + \sqrt{Z(Z+2)} \right] / M_Z + \left[\Delta IP_{1,Z}Z - \Delta IP_{1,Z+2}(Z+2) \right] / M_Z \quad . \end{aligned} \tag{66}$$

Likewise, the total energy involved in adding three electrons is:

$$\begin{aligned} &(W+H)_{\text{adding } e_1, e_2 \ \& \ e_3} = \\ &-IP_{1,1}\left[\sqrt{Z(Z+1)} + \sqrt{Z(Z+2)} + \sqrt{Z(Z+3)} \right] / M_Z + \left[\Delta IP_{1,Z}Z - \Delta IP_{1,Z+3}(Z+3) \right] / M_Z \quad . \end{aligned} \tag{67}$$

This patten generalizes to:

$$\begin{aligned} &(W+H)_{\text{adding } e_1 \text{ through } e_N} = \\ &IP_{1,1}\left[\sum_{i=1}^{N} \sqrt{Z(Z+i)} \right] / M_Z + \left[\Delta IP_{1,Z}Z - \Delta IP_{1,Z+N}(Z+N) \right] / M_Z \quad , \end{aligned} \tag{68}$$

Numerical Data to Insert in Formulae:

Numerical data for elements up to number 118 are given in [24]. The numerical analysis of EZ water requires at most the data for the first ten elements. Expressed in electron volts, eV, these numerical data are:

Hydrogen:	$Z=1$,	$M=1.008$,	$IP_{1,1}=14.250\text{eV}$,	$\Delta IP_{1,1}=0\text{eV}$.
Helium:	$Z=2$,	$M=4.003$,	$IP_{1,2}=49.875\text{eV}$,	$\Delta IP_{1,2}=35.625\text{eV}$.
Lithium:	$Z=3$,	$M=6.941$,	$IP_{1,3}=12.469\text{eV}$,	$\Delta IP_{1,3}=-1.781\text{eV}$.
Beryllium:	$Z=4$,	$M=9.012$,	$IP_{1,4}=23.327\text{eV}$,	$\Delta IP_{1,4}=9.077\text{eV}$.
Boron:	$Z=5$,	$M=10.811$,	$IP_{1,5}=17.055\text{eV}$,	$\Delta IP_{1,5}=2.805\text{eV}$.
Carbon:	$Z=6$,	$M=12.011$,	$IP_{1,6}=21.570\text{eV}$,	$\Delta IP_{1,6}=7.320\text{eV}$.
Nitrogen:	$Z=7$,	$M=14.007$,	$IP_{1,7}=27.281\text{eV}$,	$\Delta IP_{1,7}=13.031\text{eV}$.
Oxygen:	$Z=8$,	$M=15.999$,	$IP_{1,8}=27.281\text{eV}$,	$\Delta IP_{1,8}=13.031\text{eV}$.
Fluorine:	$Z=9$,	$M=18.998$,	$IP_{1,9}=34.504\text{eV}$,	$\Delta IP_{1,9}=20.254\text{eV}$.
Neon:	$Z=10$,	$M=20.180$,	$IP_{1,10}=43.641\text{eV}$,	$\Delta IP_{1,10}=29.391\text{eV}$.

Ordinary Water:

Here are some example calculations concerning possible ionic configurations of ordinary, normal water.

Most people would guess that water is $2\,H^{+}+O^{2-}$. But let us evaluate that ionic configuration. The transition $H\rightarrow H^{+}$ takes:

$$IP_{1,1}\,/\,M_1 = 14.250\,/\,1.008 = 14.1369\text{eV}. \tag{69}$$

So $2H^{+}$ takes $2\times 14.1369=28.2738$ eV.

The transition $O\rightarrow O^{2-}$ takes:

$$\begin{aligned}
&-IP_{1,1}\left[\sqrt{Z(Z+1)}+\sqrt{Z(Z+2)}\right]/M_Z+\left[\Delta IP_{1,Z}Z-\Delta IP_{1,Z+2}(Z+2)\right]/M_Z\\
&=-14.250\left[\sqrt{8\times 9}+\sqrt{8\times 10}\right]/15.999+\left[\Delta IP_{1,8}\times 8-\Delta IP_{1,10}\times 10\right]/15.999\\
&=-14.250\left[\sqrt{72}+\sqrt{80}\right]/15.999+\left[13.031\times 8-29.391\times 10\right]/15.999\\
&=-14.250\left[8.4853+8.9443\right]/15.999+\left[104.248-293.910\right]/15.999\\
&=-14.250\left[17.4296\right]/15.999+\left[-189.662\right]/15.999\\
&=-15.5242-11.8546=-27.3788\text{eV}\quad .
\end{aligned} \tag{70}$$

So the ionic configuration $2\,H^{+}+O^{2-}$ requires 28.2738−27.3788=0.8950 eV. This is a positive energy requirement, which implies that some external assistance is needed to create this ionic configuration. So normal water may *not* be $2\,H^{+}+O^{2-}$ after all!

Another possibility is readily at hand though. The ionic configuration for normal water could be $2\,H^{-}+O^{2+}$. The transition $H\rightarrow H^{-}$ takes:

$$\begin{aligned}
&-IP_{1,1}\sqrt{Z(Z+1)}\big/M_Z-\Delta IP_{1,Z+1}(Z+1)/M_Z+\Delta IP_{1,Z}Z/M_Z\\
&\quad=-IP_{1,1}\sqrt{1\times 2}\big/M_1-\Delta IP_{1,2}\times 2/M_1+\Delta IP_{1,1}\times 1/M_1\\
&=-14.250\times 1.4142/1.008-35.625\times 2/1.008+0\times 1/1.008\\
&\quad=-19.9924-70.6845+0=-90.6769\,\text{eV}\quad .
\end{aligned}\tag{71}$$

So $2\,H^{-}$ takes $2\times(-90.6769)=-181.3538$ eV. Notice that this is a *huge* negative energy. It reflects the fact that electrons *really* like to make pairs. Indeed, their propensity to do so motivated the invention of the so-called 'spin' quantum number. Without spin, many electrons in atoms would be violating the 'Pauli Exclusion Principle', which says that only *one* electron can be in any particular quantum state. Electron pairs are famous in Condensed Matter Physics too, under the name 'Cooper pairs'.

Proceeding now to the transition $O\rightarrow O^{2+}$, that takes:

$$\begin{aligned}
&IP_{1,1}\left[Z+\sqrt{Z(Z-1)}\right]\big/M_Z+\left[\Delta IP_{1,Z}Z-\Delta IP_{1,Z-2}(Z-2)\right]\big/M_Z\\
&\quad=IP_{1,1}\left[8+\sqrt{8\times 7}\right]\big/M_8+\left[\Delta IP_{1,8}\times 8-\Delta IP_{1,6}\times 6\right]\big/M_8\\
&=14.250\left[8+\sqrt{56}\right]\big/15.999+\left[13.031\times 8-7.320\times 6\right]\big/15.999\\
&\qquad=13.7907+3.7707=17.5614\,\text{eV}\quad .
\end{aligned}\tag{72}$$

Thus the creation of the water molecule in the ionic configuration $2\,H^{-}+O^{2+}$ demands altogether $-181.3538+17.5614=-163.7924$ eV. This energy is solidly negative, which means that ordinary water is overwhelmingly in this ionic configuration, $2\,H^{-}+O^{2+}$. This *is* normal water.

However, in situations where more than one version of anything can exist, both generally *do* exist, in proportions determined by their so-called Boltzmann factors, $\exp(-E/kT)$. Here E is energy, k is Boltzmann's constant, and T is absolute temperature. Boltzmann factors are the result of entropy maximization at work. Because of non-zero Boltzmann factors, there will exist a tiny, tiny fraction of the first ionic configuration, $2\ H^{+}+O^{2-}$.

This analysis of normal water shows how quantitative approaches can sometimes unseat long-standing, but never-justified, assumptions in Chemistry.

About EZ Water:

The following transition is generally thought to represent the creation of EZ water:

3 normal water molecules $\rightarrow$ 1 EZ water ion + 1 hydronium ion

That is,3(H_2O)®(H_3O_2)$^-$+(H_3O)$^+$.

Here parentheses are used to avoid implying anything about what charge the individual atoms within any ion or radical may carry. A full numerical analysis should consider all possible, or at least all plausible, ionic configurations of every molecule or radical involved.

One possible ionic configuration for the EZ water ion $(H_3O_2)^-$ is $3H^+ + 2O^{2-}$. The $3H^+$ takes total energy $3 \times IP_{1,1}/M_z = 3\times 14.250/1.008 = 42.4107$ eV. The $2O^{2-}$ takes total energy $2\times(-27.3788) = -54.7576$ eV. So the ionic configuration $3H^+ + 2O^{2-}$ altogether takes $42.4107 - 54.7576 = -12.3469$ eV. This is a negative energy, so this ionic configuration certainly can occur.

But there is also another possibility for the EZ water ion $(H_3O_2)^-$. It could have the ionic configuration $3H^- + 2O^+$. An H^- takes -90.6769 eV, so $3H^-$ takes $3\times(-90.6769) = -272.0307$ eV. An O^+ takes energy:

$$
\begin{aligned}
&(IP_{1,1}\times 8 + \Delta IP_{1,8}\times 8 - \Delta IP_{1,7}\times 7)/M_8 \\
&= (14.250\times 8 + 13.031\times 8 - 13.031\times 7)/15.999 \\
&= (114.000 + 104.248 - 91.217)/15.999 = 7.9399\text{eV} \quad .
\end{aligned} \tag{73}
$$

So $2O^+$ takes $2\times 7.9399 = 15.8798$ eV. Then the ionic configuration $3H^- + 2O^+$ takes $-272.0307 + 15.8798 = -256.1509$ eV. This energy is much more negative than that of the first candidate ionic configuration for EZ water, $3H^+ + 2O^{2-}$. This fact means EZ water is nearly always in this second candidate ionic configuration, $3H^- + 2O^+$.

One possible ionic configuration for the hydronium ion $(H_3O)^+$ is $3H^+ + O^{2-}$. This, I believe, is what most people would guess. But from the study of regular water, we know the candidate $3H^+$ would take $3\times 14.250 = 42.4107$ eV, and that the candidate O^{2-} would take -27.3788 eV, so the candidate ionic configuration $3H^+ + O^{2-}$ for hydronium would take $42.4107 - 27.3788 = 15.0319$ eV. This energy is positive, *so this ionic configuration for the hydronium ion is not promising*.

However, as was the case with the EZ water ion, there is another possibility for the hydronium ion. It $(H_3O)^+$ could have the ionic configuration $3H^- + O^{4+}$. The $3H^-$ would take $3\times(-90.6769) = -272.0307$ eV, and the O^{4+} would take:

$$
\begin{aligned}
&IP_{1,1}\left[Z + \sqrt{Z(Z-1)} + \sqrt{Z(Z-2)} + \sqrt{Z(Z-3)}\right]/M_Z + \left[\Delta IP_{1,Z}Z - \Delta IP_{1,Z-4}(Z-4)\right]/M_Z \\
&= 14.250\times\left[8 + \sqrt{56} + \sqrt{48} + \sqrt{40}\right]/15.999 + \left[13.031\times 8 - 9.077\times 4\right]/15.999 \\
&= 14.250\times 28.7361/15.999 + 67.940/15.999 \\
&= 25.5947 + 4.2465 = 29.8411\text{eV} \quad .
\end{aligned} \tag{74}
$$

So for the hydronium ion $(H_3O)^+$, the second candidate ionic configuration $3H^- + O^{4+}$ would take $-272.0307 + 29.8411 = -242.1895$ eV. This very negative energy explains why the reaction product that accompanies EZ water *is* a hydronium ion, rather than a naked proton plus a normal water molecule, which would take $14.1369 \ -163.7925 = -149.6556$eV, which is not *as* negative.

The EZ water ion and the hydronium ion together take $-268.6204 - 242.1895 = -510.8099$ eV. Compare this energy to the energy taken by three normal water molecules: $3 \times (-163.7925) = -491.3775$ eV. The EZ water ion with the hydronium ion has lower energy than the three normal water molecules. That means that Nature will take any opportunity to make EZ water ions and hydronium ions.

It appears that what creates the opportunity is a surface, plus a little energy to separate the ions. Any material body provides some gravity to create a surface, and if there is also some small energy source, such as sunlight, to help separate ions, and if there is also some normal water, the situation will automatically create EZ water too. Even an icy comet might be able to create some EZ water.

7. Microphysics

Just as the myriad compounds in Chemistry arise from not-very-many chemical elements, some significant part of the m

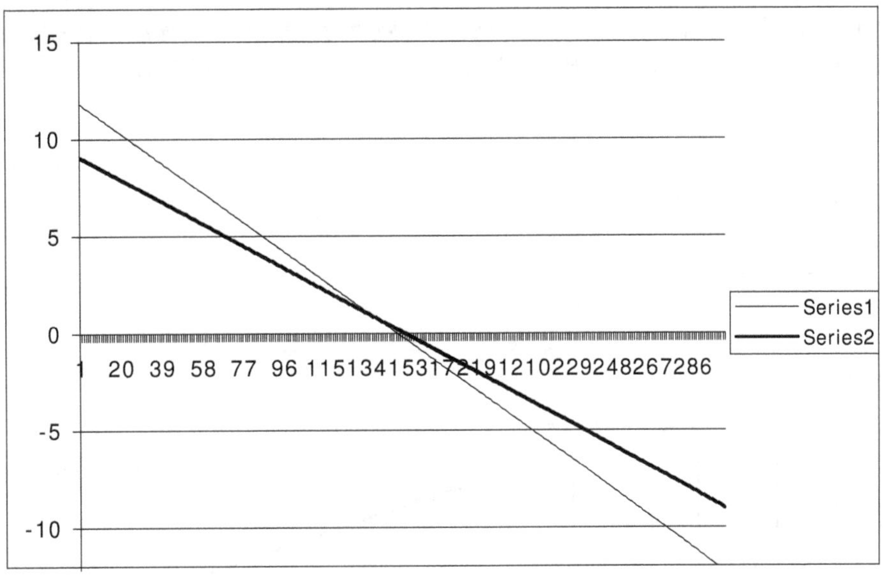

Figure 4. Log energy loss and gain rates *vs.* log system radius.

Observe that radiation dominates for radii below the crossing point, whereas torquing dominates for radii above the crossing point. This means the balance between the two effects is *unstable*: a small excursion from balance in either direction causes more excursion in the same direction. This is interesting. It means that Hydrogen does not like to exist as an isolated atom. It wants to engage in chemical reactions. In the Universe at large, you will find Hydrogen in H_2 molecules, or other molecules, or you will find Hydrogen plasma, consisting of naked protons and free electrons, but you will *not* find many isolated Hydrogen atoms.

1. We can discover even more about the Hydrogen atom if we include the appropriate angle sine and cosine factors in the calculations. These factors are oscillatory. So negative numbers sometimes occur with the torquing curve, and they cannot be plotted on a logarithmic scale. However, we are mainly interested in the points of balance between torquing and radiation, and the radiation curve is always non-negative, so the torquing curve is non-negative too at the balance points.

Figure 5 shows that we have not just one balance point, but *many* balance points. The balance points occur in close pairs, one stable and one unstable. The two pairs furthest left on the plot do not show as crossings because of the finite resolution of the plot, but they are certainly present.

More solution pairs are to be found, off the Figure to the left. Indeed, the solution pairs continue *indefinitely*, into smaller and smaller system radii. So Hydrogen has an infinite family of 'sub-states'. Mills discusses these in [27].

The smaller and smaller radii of the balance points in Fig. 5 correspond to higher and higher orbit speeds. This idea conflicts with a prohibition imposed by SRT: no physical particle possessing mass is allowed to move at a speed matching or exceeding light speed c. What does this conflict mean? I believe it means the prohibition should be understood more precisely to say: no physical particle possessing mass *can be* **perceived** to move at a speed matching or exceeding light speed c, **if** *we agree to process all received data in accord with Einstein's Second Postulate.* If we do not agree to that, then there is no particle speed limit.

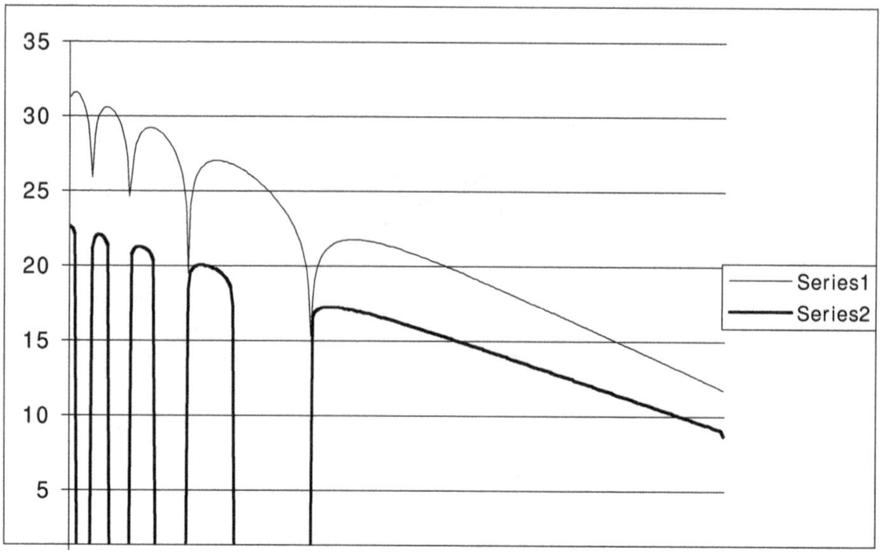

Figure 5. More and more balance points below the Hydrogen ground state.

2. We can also study positronium (a system consisting of one electron and one positron). See Fig. 6. As with Hydrogen, the oscillatory angle factors create a family of solutions for this system too. Half of them are stable, and half are unstable, and uncountably many of them occur at small radii and high speeds well in excess of c. Due to the finite resolution of the plot, only one pair of solutions is clearly visible in Fig. 6. But two more, at smaller radius and higher speed, are also certainly present. We can characterize these, and all high-speed solutions, without even knowing exactly what the radiation curve is like - its exact amplitude, or its r^{-4} dependence. The one low-speed solution is just $v_0 \approx 0.02 \times c$. The many high-speed solutions have to occur in pairs just above and below orbit speeds of the form $v_n = v_0 + n \times 2\pi c$, where n is an arbitrary positive integer.

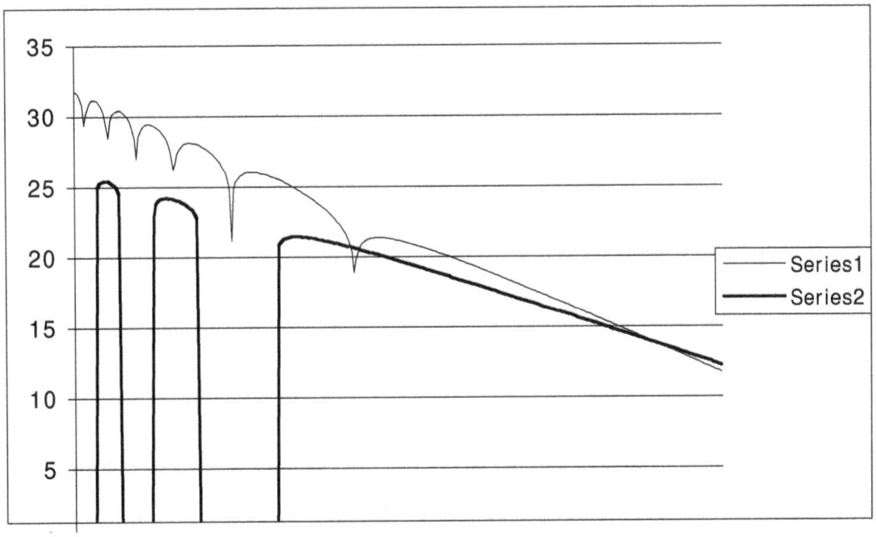

Figure 6. Positronium solutions.

A parenthetical note applies for Fig. 6: for same-mass systems, the amplitude of the radiation curve should be less by a factor of 4 because there is no center-of-mass motion. This sort of numerical detail does not significantly affect where the solutions fall. That is determined almost entirely by the cosine factors that produce the deep dips that intersect the peaks in the curve for rate of energy gain due torquing.

3. The oscillatory nature of the angle factors can turn a situation of seeming repulsion into a situation of actual attraction. This phenomenon of sign reversal due to signal delay is well known to engineers, who often deal with oscillating signals in feedback control systems For the present application, consider two electrons in a circular orbit, and suppose they move at speed πc. One electron launches its signal radially outward. By the time this electron has executed half an orbit, this signal has expanded a distance equal to the orbit diameter. By then, the two electrons have exchanged places. So the expanding signal first contacts the second electron at exactly the signal launch point. Then the two electrons complete their orbit. At the end, the second electron finally understands its signal: it is to move radially. But by now, the two electrons have changed places again, and for the second electron, the direction commanded is *inward*. That situation is equivalent to *attraction*.

Given this mechanism for attraction, we can also study homogeneous systems: two electrons, or two positrons, for example. Again, there exist both stable and unstable solutions, and there are infinitely many of each, corresponding to orbit speeds of the form $v_n = \pi c + n \times 2\pi c$ for arbitrary positive integer n. See Fig. 7.

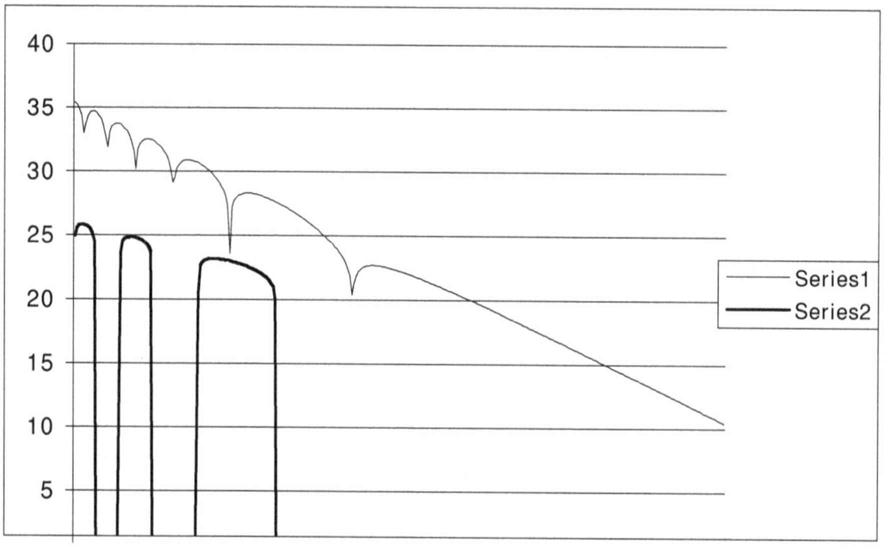

Figure 7. Same-charge solutions.

An additional parenthetical note applies for Fig. 7: for same-charge systems, the angular pattern of the radiation is quadrupole, rather than dipole, so the amplitude of the radiation energy loss curve should decline as r^{-6} rather than r^{-4}. Again, this sort of numerical detail does not significantly affect where the solutions fall, which is determined almost entirely by the cosine factors that produce the deep dips that intersect the peaks in the curve for rate of energy gain due torquing.

The stable solutions for two electrons bring to mind the situation that is so well known in Chemistry: electron pairs. They are everywhere in Chemistry. The most famous case occurs for Helium. Helium is a noble gas, and it reacts with other elements only under extreme duress. Helium has two electrons, and pulling one electron away is very costly: Helium has the highest ionization potential of any element. The message is: two electrons definitely do form a stable subsystem within an atom. The standard QM explanation for this invokes the concept of electron spin, with two possible values, $\pm\hbar/2$, allowing two electrons in the same overall energy state. Electron pairing also occurs famously in Solid State Physics, under the name of Cooper Pairs.

8. Conclusion

In the present Chapter, the new concept applied is the more realistic signal model for use in an improved version of SRT. The realistic signal model is based on Information Theory,

The new concept is implemented with very standard mathematics: differential equations, their family of solutions, and the particular problem boundary conditions. These mathematical ingredients for a proper signal model were all available in 1905, but they were not used in SRT. Why? I believe the fundamental reason is the history: Information Theory was not yet available, so no researcher at that time would have been likely to detect the inadequacy of the infinite plane wave as a signal model.

The present paper has shown that there are rewards for instead using the realistic photon/ signal model. They include more insight into Quantum Mechanics, and into Gravity Theory, and potentially into Elementary Particle Physics. These are all subjects to be studied much more fully in the future.

Many textbook treatments of SRT devote a lot of space to Lorentz Transformations (LT's). The present work has not mentioned LT's at all. To this author, LT's just seem to describe the wrongly informed opinions of different observers. So I don't really want to focus on LT's. But I have to mention them, because repairing SRT to take proper account of the concepts of IT casts doubt on Einstein's SRT, and hence on LT's. Therefore, I hereby relegate the unavoidable discussion of coordinate transformations, LT's and others, to the following Appendix.

9. Appendix

The situation in the late nineteenth century included the following fact: Maxwell's first order coupled field equations appeared *not* to be invariant under Galilean transformation of coordinates (GT's). Phipps [28] has written extensively about this apparent conflict between Maxwell's Electromagnetic Theory and Newton's Mechanics. In the early twentieth century, SRT brought in LT's, and the conflict seemed to be resolved: Maxwell's electromagnetic theory was clearly invariant under LT's. This fact was taken as evidence in favor of Einstein's SRT over Newton's Mechanics.

But there is a puzzle left to resolve: Maxwell's first order coupled field equations appear to qualify as *tensor* equations. Mathematicians had developed tensors in the first place to enable the articulation of mathematical statements that would be *coordinate-free*. So tensor equations are *by definition* invariant to *all* invertible coordinate transformations.

So what had happened here? I believe two circumstances had collided to create a very bad situation. One circumstance was that Mathematics had such a long history of developing *eternal* truths: the focus had been on arithmetic, geometry, and trigonometry – all of them eternal in character. Even archeo-astronomy was largely about the eternal *repetition* of events, and *not* about temporal *evolution* of events. Eternal truths really need not have a time dimension. They can, however, have as many spatial dimensions as may be desired, and that became the focus for much of tensor analysis. The other circumstance was that time became a really significant variable with the advent of modern Physics: Kepler, Galileo, Newton, and Maxwell. And time is a really different kind of variable than space is. Maxwell was very well aware of the difference, as he developed his electromagnetic theory in terms of Hamilton's quaternions. The modern equivalent of the quaternion tool is the set of four 2×2 complex Pauli spin matrices:

$$\sigma_{ct}=\begin{bmatrix}1&0\\0&1\end{bmatrix},\sigma_x=\begin{bmatrix}0&1\\1&0\end{bmatrix},\sigma_y=\begin{bmatrix}0&-i\\i&0\end{bmatrix},\sigma_z=\begin{bmatrix}1&0\\0&-1\end{bmatrix}. \tag{75}$$

The first one, the time-like one, is the identity matrix. The other three, the space-like ones, produce the identity matrix when squared. When two of them are cross-multiplied, they generate a factor of $i=\sqrt{-1}$ times the third one, corresponding to the vector cross product in three-dimensional space.

The collision between circumstances came in the formulation of differential operators. People were familiar with the scalar chain rule,$\boldsymbol{d/dt}=\partial/\partial_t+v\partial/\partial_x$, and did not realize that more information was needed in the context of vector and tensor applications with time as well as space dimensions.
The 2×2 Pauli matrices are easy to appreciate visually, so I will also discuss transformations of coordinates in terms of 2×2 matrices also - but only real ones, not complex ones. Let s stand for any spatial coordinate. A general coordinate transformation involving ct and s has the form:

$$\begin{bmatrix}ct'\\s'\end{bmatrix}=\frac{1}{\sqrt{1-AB}}\begin{bmatrix}1&B\\A&1\end{bmatrix}\begin{bmatrix}ct\\s\end{bmatrix}. \tag{76}$$

For the familiar Lorentz transformation, $A=B=-v/c$, where v is the speed of the new coordinate frame relative to the old one. The letter v is lower case to remind us that $v<c$; *i.e.* $v/c<1$. We have:

$$\begin{bmatrix}ct'\\s\end{bmatrix}=\frac{1}{\sqrt{1-v^2/c^2}}\begin{bmatrix}1&-v/c\\-v/c&1\end{bmatrix}\begin{bmatrix}ct\\s\end{bmatrix}. \tag{77}$$

For the long discarded Galilean transformation, $A=-V/c$ and $B=0$. The letter V is upper case to remind us that V is *not* limited, and might exceed c. So we have:

$$\begin{bmatrix}ct'\\s\end{bmatrix}=\frac{1}{1}\begin{bmatrix}1&0\\-V/c&1\end{bmatrix}\begin{bmatrix}ct\\s\end{bmatrix}. \tag{78}$$

For all such general coordinate transformations, there also exists a complement transformation:

$$\begin{bmatrix}ct''\\-s''\end{bmatrix}=\frac{1}{\sqrt{1-AB}}\begin{bmatrix}1&-A\\-B&1\end{bmatrix}\begin{bmatrix}ct\\-s\end{bmatrix}. \tag{79}$$

Its purpose is to preserve inner products; for example:

$$\begin{bmatrix} ct & s \end{bmatrix} \begin{bmatrix} ct \\ -s \end{bmatrix} = (ct)^2 - s^2. \tag{80}$$

Observe that:

$$\begin{aligned} \begin{bmatrix} ct' & s' \end{bmatrix} \begin{bmatrix} ct'' \\ -s'' \end{bmatrix} &= \begin{bmatrix} ct & s \end{bmatrix} \frac{1}{\sqrt{1-AB}} \begin{bmatrix} 1 & A \\ B & 1 \end{bmatrix} \frac{1}{\sqrt{1-AB}} \begin{bmatrix} 1 & -A \\ -B & 1 \end{bmatrix} \begin{bmatrix} ct \\ -s \end{bmatrix} \\ &= \begin{bmatrix} ct & s \end{bmatrix} \frac{1}{1-AB} \begin{bmatrix} 1-AB & 0 \\ 0 & -BA+1 \end{bmatrix} \begin{bmatrix} ct \\ -s \end{bmatrix} \equiv (ct)^2 - s^2 \quad . \end{aligned} \tag{81}$$

For Lorentz transformation, the complement transformation is the inverse, or equivalently, the *re*verse transformation:

$$\begin{bmatrix} ct'' \\ -s'' \end{bmatrix} = \frac{1}{\sqrt{1 - v^2 / c^2}} \begin{bmatrix} 1 & +v/c \\ +v/c & 1 \end{bmatrix} \begin{bmatrix} ct \\ -s \end{bmatrix}. \tag{82}$$

But for Galilean transformation, the complement transformation is:

$$\begin{bmatrix} ct'' \\ -s'' \end{bmatrix} = \frac{1}{1} \begin{bmatrix} 1 & +V/c \\ 0 & 1 \end{bmatrix} \begin{bmatrix} ct \\ -s \end{bmatrix}. \tag{83}$$

This *is* the *in*verse transformation, but *not* the *re*verse transformation. It is rather the *transpose* of the reverse transformation. It looks so very strange because, for more than a century now, *only* Lorentz transformations of velocity have been used in mainstream theoretical Physics, and transposition does not change them.

The vital role for this strange new thing lies with the differential operators. The story is much like it was for the coordinates: there are *two* complementing transformations, and they involve, not only inversion/reversal, but also transposition. Let ∂_{ct} represent differentiation with respect to the time-like coordinate, and ∂_s represent differentiation with respect to the spatial variable. Let us demand invariance of inner products involving differential operators; for example, like:

$$\begin{aligned} &\begin{bmatrix} \partial_{ct} & \partial_s \end{bmatrix} \begin{bmatrix} ct \\ -s \end{bmatrix} = 1 - 1 = 0 \text{ and } \begin{bmatrix} \partial_{ct} & -\partial_s \end{bmatrix} \begin{bmatrix} ct \\ s \end{bmatrix} = 1 - 1 = 0 \\ &\text{or } \begin{bmatrix} \partial_{ct} & \partial_s \end{bmatrix} \begin{bmatrix} \partial_{ct} \\ -\partial_s \end{bmatrix} = \partial_{ct}^2 - \partial_s^2 \text{ and } = \begin{bmatrix} \partial_{ct} & -\partial_s \end{bmatrix} \begin{bmatrix} \partial_{ct} \\ \partial_s \end{bmatrix} = \partial_{ct}^2 - \partial_s^2. \end{aligned} \tag{84}$$

i.e., always *two* statements – not just *one* statement. This level of detail was missing from the *scalar* chain rule, and that omission caused people to believe that Maxwell's equations could not be shown to be invariant under GT. And so they welcomed LT instead. This is not to say we should now revert to using GT again. Indeed, because of the half-retardation issue discussed in Sect. 3, the best transformation to use may involve, not V / c, but rather $V / \underline{2}c$. This question needs detailed future study.

The use of 2×2 matrices can make the detail needed in such future study very clear. However, many mathematicians tend to prefer tensor notation. But current-day tensor notation uses only two index positions, both on the right: down called 'covariant', up, called 'contravariant'. To represent the transformations needed for Physics, it would be helpful, and maybe necessary, to add two more index locations, up and down on the *left*, to acknowledge transposition, and using words like 'trans-covariant' and 'trans-contravariant' to emphasize what putting indices in those positions means.

Acknowledgements

This Chapter is dedicated to the memory of a most courageous researcher in theoretical and applied electrodynamics: Dr. Peter Graneau, 1921-2014. He encouraged me, and many other researchers, to give serious attention to the History behind Physics.

Author details

Cynthia Kolb Whitney*

Address all correspondence to: galilean_electrodynamics@comcast.net

Galilean Electrodynamics, USA

References

[1] Einstein, A., On Thr Electrodynamics of Moving Bodies, Annelen der Physik 17, 891-921 (1905).

[2] Einstein, A., Annelen der Physik On the Theory of Light Production and Light Absorption, Annelen der Physik 20, 199-206 (1906).

[3] Brillouin, L., *Science and Information Theory*, Second Edition, Dover Publications, Inc., Mineola, NY, (2013).

[4] Flores-Gallegos, N., Shannon Informational Entropies and Chemical Reactivity, Chapt. 29, pp. 683-722 in *Advances in Quantum Mechanics,* Intech (2013).

[5] Jackson, J.D., *Classical Electrodynmics,* Second Edition, John Wiley & Sons, New York, NY. (1975).

[6] Bentwich, J., The Theoretical Ramifications of the Computational Unified Field Theory, Chapt. 28, pp. 671-681, *Advances in Quantum Mechanics,* Intech (2013).

[7] Whitney, C.K., Better Unification for Physics in General Through Quantum Mechanics in Particular, Chapter 7, pp. 127-160 in *Theoretical Concepts of Quantum Mechanics,* Ed. M.R. Pahlavani, InTech. (2012).

[8] Ritz, W., Researches critiques sur l'electrodynamique generale, Ann. Chim. et Phys. 13, pp. 145-275, (1908).

[9] Putz, M.V., Path Integrals for Electronic Densities, Reactivity Indices, and Localization Functions in Quantum Systems, *International Journal of Molecular Sciences* 10, pp. 4816-4940 (2009).

[10] Bracken, P., Quantum Mechanics Entropy and a Quantum Version of the H-Theorem, Chapt. 21, pp. 469-488 in *Theoretical Concepts of Quamtum* Mechanics 469-488 (2012).

[11] Bracken, P., The Schwinger Action Principle and its Applications to Quantum Mechanics, Chapt. 8, pp 159-182 in *Advances in Quantum Mechanics,* Intech (2013).

[12] Spitnev, V., Generalized Path Intergral Technique: Nanoparticles Incident on a Slit Grating, Matter Wave Interference, Chapt 9., pp. 183-212 in *Advances in Quantum Mechanics,* Intech (2013).

[13] Bulnes, F., Correction, Restoration and Re-Composition if Quantum Mechanical Fields of Particles by Path Integrals and Their Applications, 489-514 in *Advances in Quantum Mechanics,* Intech (2013).

[14] Liénard, A., Champ Electrique et Magnétique produit par une Charge Electrique Concentrée en un Point et Animée d'un Movement Quelconque, *L'Eclairage Electrique, XVI,* pp. 5-14, 53-59, and 106-112 (1898).

[15] Wiechert, E., Elektrodynamische Elementargesetze, *Archives Néerlandesises des Sciences Exactes et Naturelles, série II, Tome IV,* pp. 549 – 573 (1901).

[16] Lokajicek, M., V. Kundrát, and J. Procházka, Schrödinger Equation and (Future) Quantum Physics, Chapter 6 in Advances in Quantum Mechanics, Ed. Paul Bracken, InTech. (2013).

[17] Whitney, C.K., On the Dual Concepts of 'Quantum State' and 'Quantum Process', Chapter 17 in *Advances in Quantum Mechanics,* Ed. Paul Bracken, InTech. (2013).

[18] Enders, P., Quantization as Selection Rather than Eigenvalue Problem, Chapt 23, pp. 543-564, *Advances in Quantum Mechanics,* Intech (2013).

[19] Ralston, J.P., Emergent un-Qusntum Mechanics, Chapt. 19, pp. 437-475, in *Advances in Quantum Mechanics*, Intech (2013)

[20] Whitney, C.K., Reasonable doubt: Cosmology's gift for physics, *Journal of Computational Methods in Science and Engineering* 2013, *13*, pp. 291-302 (2013).

[21] Graneau, P., *Ampère–Neumann Electrodynamics of Metals*, Hadronic Press, Nonantum, MA, USA, pp. 7 – 22 (1987).

[22] MacGregor, M., *The Enigmatic Electron- A Doorway to Particle Masses*, 2nd ed; El Mac Books, Santa Cruz, CA, USA; pp. 3 –11 (2013).

[23] Pollack, G., *The Fourth Phase of Water, Beyond Solid Liquid Vapor*; Ebner & Sons, Seattle (2013).

[24] Whitney, C.K., *Algebraic Chemistry: Applications and Origins*, Nova Science Publishers (2013).

[25] Whitney, C.K., EZ Water: A New Quantitative Approach Applied – Many Results Obtained, Water 5, January 5, pp. 105-120 (2014).

[26] Whitney, C.K., The Neutron: a Challenge for Post-Maxwell Physics, *Hadronic Journal*, 3, pp. 201 – 250 (2008).

[27] Mills, R.L., The Nature of the Chemical Bond Revisited and an Alternative Maxwellian Approach, *Physics Essays* 17, 3, pp. 342 – 389 (2004).

[28] Phipps, T.E., *Heretical Verities: Mathematical Themes in Physical Description*; classic non-fiction library, Urbana, IL, USA, pp. 102 – 108 (1986).

6

Unification of Quantum Mechanics with the Relativity Theory, Based on Discrete Conservations of Energy and Gravity

Aghaddin Mamedov

1. Introduction

The history of physics has two great revolutionary theories, such as relativity and quantum physics. However, the new discoveries of the particle physics do not fit with the principles of both theories. One of the main questions is how to explain the break of symmetry in proton-antiproton collision experiments and formation of more matter particles than that of anti-matter ingredients. The theory, which can explain this phenomenon, should answer to the question how matter in the beginning of universe formed and what is the space-time structure of the universe.

In accordance with the modern physics, the small-scale experiments of particles physics have to be described by the Standard Model of quantum mechanics. However, the phenomenon of matter/antimatter symmetry breaking appearing at subatomic scale requires formulation of the new dynamical laws. Presently it is not clear that the mystery of the small-scale dynamics is due to the incompleteness of the mathematic formulation of dynamics of physical events at small scale or to the change of mathematics. The problem is that our present knowledge on mathematical description of change at small scale of space-time frame and physical theories do not distinguish what special features should have the initial state (position) of universe.

Quantum mechanics suggests that at sub –atomic scale the features of initial sate, the position and velocity of a particle cannot be measured. At this scale, the differential equations of dynamical evolution do not work and the change of the state of a system cannot be found by its velocity and position where takes place appearance of wave-like performance.

The problem is that the concept of "rest energy" or "rest mass" can not be explained on the basis of the our present knowledge of static energy conservation law. The static state of "rest mass" requires application of uncertain amount of energy to keep a body at certain constant position within infinite time duration.

Quantum mechanics relates the problem of "rest" to the uncertainty principle. By quantum mechanics, a particle cannot be at rest because at the "rest" the position and velocity both have to be certain, which cannot be possible.

In classic formulation of Neother theorem, the concept of energy conservation is uncertain due to the continuous feature of the energy conservation. The symmetry principle, revealed from this theory became also uncertain. Therefore, the scientific basis of symmetry breaking, observed at small scale, cannot be explained because of classic energy conservation principle.

The energy as an identity may be conserved discretely because continuous energy conservation in the form "energy can not be destroyed and created" leads to the arbitrary and infinite feature of the dynamical events: the amount and origin of conserved energy within this formulation is uncertain.

The non-continuous energy conservation in the non-arbitrary frame requires existence of boundary of the energy conservation, which has to be localized within space-time framework. But our present understanding of space-time frame does not describe what the space and time are.

There are many concepts, related to the nature of space and time variables but the true nature of time (space as well) is not known. Particularly it is not clear why space has tree dimensions while time has only one. Due to this reason all the physical theories, describing motion distinguish difference between the past and future. The second law of thermodynamics, correlating irreversible time arrow with the increase of randomness does not explain evolution of biological events in direction of more ordered states.

By relativity, space and time are not absolute variables.Unfortunately, the theory of relativity is the theory of geometric space-time and it does not explain why the space-time unit connects the parameters having different dimensions.

Based on these problems, the aim of our chapter is to give analysis of the quantum mechanics and relativity within the principles of classic energy conservation law and to describe main features of these theories within alternative concept of discrete energy conservation. In the frame of discrete energy conservation the classic symmetry principles gets entirely new feature which occupies the special part of the present work.

In the present chapter, we also discuss one of the mysteries of the physics related to the transformation of the space-time variables. Physics accepts invariance in time, which does not fit with the CP violation. We will discuss also Paolo Scaruffi analysis [1] on the laws of classic physics that "why do objects in accordance with the first law of motion have a preference for travel in a straight line at constant speed is not clear. Where does this property come from? It is not clear also, why a body in motion tends to remain in motion and a body at rest tends to say at rest. These questions are open to interpretation".

In our work, we have suggested a new space-time dynamic boundary mapped discrete energy/ momentum conservation law for unification of quantum mechanics and relativity scales interactions, which is based on a new non-Lagrangian mathematical foundation of non-invariance action principle and commutation of frequency domain differential operators of conjugated canonical space-time variables. We showed that the phenomenon called space is the materialization of background "vacuum" energy in the momentum of dark matter, while time phase destroys everything material, reversing the alignment of dark matter momentum to the initial state of "vacuum" energy. Therefore gravity is not the geometry of space-time, it is the dynamic order to hold discrete energy conservation cycles and its power source is the background coupling of space-time phases.

The mechanism of mass generation has been described in the chapter in detail. It is shown that generation of correlated mass and its conservation takes origin by consumption of energy of discrete space-time background frame, moving in direction of space phase expansion. The "continuous parity" conservation may result only by discrete commutation of **SU (2)** group of interactions with the coupling energy of background virtual space-time ingredients, recycling its discrete uniform energy to the **2:1** confined interactions of nuclear. The uniform exchange of energy, generated by not-yet observed background **[T-B]** bi-meson field through interaction with the **uu**ˉquarks, regulates discrete energy conservation within nuclear and background polarization state.

The new non-Lagrangian time/space dependent function shows that the "resulting weak response force" gets its origin from the action-advanced response non-invariant parity of three-component interactions **(V-A)/A,** providing a new deterministic description of nature and fundamental dynamic symmetry. With the static, single non-correlated existence of the identity and physical parameters, the energy and momentum (mass) are not conserved. The physical reality and its parameters as the resulting (not passive relativistic) quantities, do not exist independently of the advanced response interactions: in formulations of classic and quantum physics, the fundamental Lagrangian action is not conserved.

2. The physics of Einstein's relativity

The question "What is the wrong with the relativity theory?" appears frequently in discussions related to the relativity theory. Of course, the relativity theory is not a complete theory. Particularly, the energy –matter equivalence in $E=mc^2$ formulation is not complete from two reasons: the concept of rest mass and the conditions of energy-mass equivalence of the special relativity is not clear. What mass is equivalent to the energy and what features has the rest mass as an equivalent to the energy? What is the performance of this equivalency at dynamical conditions out of the vacuum? What is the connection of the mass with the dynamics of the space-time frame? There is no physical meaning of the increase of the inertia (mass) with the increase of the velocity even with application of Lorentz transformations.

The special statement of relativity that space and time variables are not absolute variables and they form the non-separable space-time unit is very important statement of this theory. From this statement follows that description of the change by classic differentiation where change of the function (space function, such as length) has a relation with the independent time input leads to the appearance of the uncertainty. Therefore, change of space has to be correlated with the response of the time to the change. Unfortunately, this concept, which follows from the analysis of relativity, was not realized in the mathematic formulation of the special relativity.

General relativity is not a complete theory because it cannot describe space-time structure at small scale. However, general relativity, as the special relativity, also has an important statement that space-time structure is determined by the energy and mass in it, which makes it non-linear. The problem is that general relativity describes this relation by geometry, which cannot show change of the structure of space-time with the change of energy and matter. Due to this incompleteness, the general relativity leads to the singularity at small-scale physics (atomic scale) and at high space scale called black holes.

Another big problem of the relativity is its concept, based on continuous dynamics, which leads to the relation of the motion to the relative reference frames. Elimination of continuity and description of the natural events by discrete dynamics do not need application of "reference frames" for relation of the motion. In discrete dynamics, the motion is related to the action force: in the presence of the action, an event changes its state, but when there is no action force an event returns to its initial state.

Based on the discrete dynamics, the Newton's laws of motion and gravitation could be unified. The motion is the result of the action force, therefore the Newton's first law that' if a body is not affected by any force it will keep its initial uniform motion in straight line" eliminates that the body before "affect of any force" had a motion due to the result of some force.

Einstein showed that space and time are simply different dimensions of the same space-time continuum. All events and quantities decompose into time and space components, which depend on the observers. Einstein connected the curvature of space-time of an event with the energy and momentum of the objects. By Einstein opinion the energy and momentum are the same quantities of space-time which has four dimensions. That is why space-time is the same in relation for all reference frames and change of the event is realized through change of the space and time components of this frame. The relative quantity of energy and momentum depends on the observer.

The problem of this approach is that the dynamical space-time variables were connected within continuum framework, which did not allow distinguish personal properties of time and space identities. Later we will show that the difference in performance of space and time variables do not involve dimension of these parameters but deeply is connected with the asymmetric phases, binding the boundary of the space-time phases. The relative quantity of energy and momentum has to be determined by the asymmetry in the boundary of these variables. The contraction of the length and time dilation cannot be described without specification of the boundary of these variables. Therefore, contraction of space and time delay cannot be without relation of these parameters to their initial boundary. The relative ratio of space-time variables in the dynamic space-time frame (by relativity-decomposition of space-time continuum to its ingredients) may be different for the observers, participating differently in the space-time event and sharing directly or indirect the energy recourses with the event.

In Newtonian physics, time flows at a constant rate for all observers. This statement of Newtonian physics appears from the concept of "independent time". Therefore, in accordance with the Einstein's relativity flow of time has different rate for different observers. Therefore, by relativity time is the personal property of an event: different observers, giving different time flow for the same event, present their "personal time". The problem is that relativity equations do not clarify why different observers could measure the same speed of an event by their own measurements. It is not known also how to unify the measurements of the different observers.

Einstein determines the dynamics of matter by the geometry of space-time and that geometry is determined by the distribution of matter. As in the case of special relativity, in general relativity many questions remain open. Particularly general relativity does not explain the origin of mass and energy, which curves the structure of space-time. The question related to the Newton physics, how the moving body responses to the action, in the relativity theory also remains open. If this energy has an external origin, how this energy is generated and how it interacts with the space-time remains open. But if this energy has an internal origin within space-time frame, the mechanism of the energy generation and the energetic features of space is the subject of interpretation. It is not clear also if the universe is a single system, what energy makes its expansion.

The main question here is the energy content of the space. What is the property of the vacuum, is it the empty unit? How the empty space may have energy? Unfortunately, the basic formulations of special and general relativity do not provide answer to these questions without which the relativity theory itself became the "observer" between Newton's physics and quantum mechanics.

3. Fundamental basis of quantum principles and their problems

Quantum mechanics did not solve problem of classic mechanics by modification of its causal dynamical laws but applied entirely new concept of probability, which was a new philosophy for description of nature. Description of the causal nature by mathematical formulation of classic differentiation had boundary problems, which lead to the uncertainty [2]. Description of the position and velocity by classic approach through application of boundary of variables to the mathematical formulation could solve the problems. Dirac's relativistic quantum mechanics concept is the realization one of these ways but Dirac's concept kept the probability feature of the quantum mechanics.

The fundamental basis of the quantum mechanics was very important due to the application the uncertainty in the description of the "change phenomenon" instead of mathematic formulation of classic physics. Unfortunately, the derivative of the function without involving of the function itself leads to the uncertainty. Therefore, the physical observables, such as momentum and velocity of the classic formulation were replaced by the possible states of operators on a wave function. The starting point of quantum mechanics was that it transformed the physical parameters to the differential operators: the time operator replaced energy while momentum by spatial operator.

In accordance with the quantum mechanics, the physical state of a body cannot be described by classic state and it has to be described by "quantum state", presented by wave function. The change of the wave function describes evolving of a system with time. It is clear that during change of a system with time the evolving wave function cannot be merged with the law of relativity.

It is necessary to note that Schrödinger equation is not free from the continuous energy conservation principle due to the involvement of single time derivative as an energy operator.The another problem of the Schrödinger equation is that it involves Hamiltonian as the linear operator of the wave function. The Schrödinger equation is a single particle's equation that is why Hamiltonian does not describe the interaction of the particle with its surrounding medium. Due to this features this equation has a problem of "locality" of a body, described by the wave function.

Due to the absence of "absolute frame of reference", description of the state of a particle was one of the difficult subjects of the relativistic and non-relativistic concepts. Shrodinger equation contains only first order derivative with respect to the time. Dirac tried to give description, such as a differential equation first order for space and time, which may lead to the simple relativistic relation between energy and momentum.

To eliminate the problems of the Schrödinger equation, Dirac suggested the relativistic quantum mechanical wave function in order to fit the concept of relativity. Dirac equation is the first attempt for generation of theory consistent with the principles of quantum mechanics and relativity theory.

Dirac equation was similar to the Schrödinger equation, but his proposal suggested existence of the anti-matter. The main principle of Dirac equation [3, 4] is that Hamiltonian, correlated with the input wave function, was described through complex value, involving the sum of energy-momentum consistent:

$$\left(\beta mc^2 + \sum_{k=1}^{3} \alpha_k p_k c\right)\psi(\mathbf{x},t) = i\hbar \frac{\partial \psi(\mathbf{x},t)}{\partial t} \tag{1}$$

Dirac assumed that presenting Hamiltonian through energy-momentum sum might describe the atomic spectra and discrete angular momentum of an electron.

It is clear that the form of the wave function and its evolution is determined how the total energy and energy-momentum has been described. Hamiltonian of Schrödinger wave function involves the total energy of a system. Making the Schrödinger equation relativistic, Dirac used some additional transformation expressions and got a equation

$$-\frac{\hbar^2}{2m}\nabla^2\phi = i\hbar\frac{\partial}{\partial t}\phi. \tag{2}$$

Dirac equation in this form still is not a complete dynamical model because it presents the dynamic behavior of a free particle and cannot explain why an electron has a spin angular momentum of half a quantum. The Dirac's relativistic wave equation [3, 4] was the first try to communicate relativity with the quantum mechanics.The left side of the equation (1) describes momentum ingredients:

$$\left(\beta mc^2 + c\left(\alpha_1 p_1 + \alpha_2 p_2 + \alpha_3 p_3\right)\right)\psi\left(x,t\right) = i\hbar\frac{\partial\psi\left(x,t\right)}{\partial t} \tag{3}$$

It is necessary to note that Dirac proposal, suggesting existence of anti-matter (particularly positron) does not explain properly physical state of negative energy, introducing it as a "sea" of negative particles. It could not explain also how matter may degenerate into negative-energy states.

The relativistic feature of Dirac's proposal was explained by his suggestion that relativistic description of a particle should involve multiple wave functions for other potential particles.

But Dirac's equation does not involve first order space-time derivatives which in the form of symmetric ingredients have to describe energy and momentum parts of the space-time frame. Dirac's equation also does not involve any information about initial values of the wave function at every discrete time instants; therefore, it could not lead to the definite solutions for communication of relativity with the quantum mechanics. The initial values of wave function of Dirac equation are freely chosen.

The special feature of Dirac's concept is the "Dirac sea" which is related to the negative energy solutions.The "negative energy" of Dirac's sea is explained with the principle that every quantum state can only be occupied with one electron. This concept accepts that the total energy and total charge are infinite. Therefore, this principle does not explain the source of infinite energy and infinite charge.The Dirac's sea is possible only with the infinitely many particles.

In accordance with the Dirac concept, when one electron is lifted form "Dirac's sea" via a high energy of γ-particle, there forms "a hole in the Dirac sea". By this concept, the "hole" represents the anti-particle to the electron with the positive charge, namely the positron, which has been proven experimentally.

Dirac equation can be obtained from the lagranjian action principle, but lagranjian action cannot give first order derivatives for the event simultaneously in space and time variables.

O.Klein and W.Gordon proposed [5, 6] the concept, which is applied to describe "particle" behavior in the relativistic mode and an equation, which has to simulate the behavior of a spineless free particle which has a spinless wave solutions. The Klein-Gordon equation is a relativistic version of Shrodinger equation, it is the second order in time, and it describes spinless particles in towards and backwards in time.

The Klein–Gordon equation has the following description:

$$\frac{1}{c^2}\frac{\partial^2}{\partial t^2}\psi - \nabla^2\psi + \frac{m^2c^2}{\hbar^2}\psi = 0. \quad (4)$$

The Klein-Gordon equation is the second order for time expression and by separation of positive and negative parts describes time-independent case. In some versions, Klein-Gordon equation was introduced with the second order for time and space but in these expressions have boundary value problems.

Although Klein-Gordon equation claims to be relativistic equation, it does not give complete relativistic picture of a "particle" dynamics simultaneously in space and time coordinates, which in relativistic mode needs to be connected with the space-time frame, required by relativity theory.

4. Discrete conservation of energy as the basis for unification of physics

4.1. Formulation of discrete space-time field theory

The concept, which we are planning to use, has to eliminate the small-scale phenomenon-point particle concept of classic physics and quantum mechanics probability of location of subatomic particles. The new concept is the discrete dynamical structure of space-time phases. The position of an elementary particle, located within space-time phases is not a point; it exists as a time carrying identity within minimum space frame, called elementary space-time manifold commuting space and time phases.

The problem of classic physics is the description of the events by displacement of a point in space (**Δx**) within certain displacement of time (**Δt**), presented in the form of relative simple intervals. Two point particles having the same displacement in space and time could pass the observer's reference position differently being in "phase" or in "opposite phase" due to the having, phase differences starting from different initial conditions.

Feynman [7] showed that elimination of the infinity in dynamical formulations could be done through substitution of interval of function (**Δf**) by the value of the function itself (**f**) but by his opinion in this case the dynamical event will be static.

It is clear that the effect of scale phenomenon to the behavior of dynamical systems can be analyzed by Hamilton canonical coordinate transformations. Unfortunately, a genetic Hamilton's transformation (**g, p, t**) → (**Q, P, T**) has no explicit time dependence and does not preserve this transformation.

We used a new principle of canonical transformation where the transform is the time – frequency representation. The new approach involves commutation of frequency with the time domain through conjugation of the change of function (**Δf**) with the local function (**f_n**) itself. In this case, the reciprocal discrete transform within (Δf) and function (f_n) itself can be generalized to the Abelian group due to the generation of dynamic translation within boundary of canonical variables.

We did not replace the interval of the function (**Δf**) with the function itself (**f_n**), as Feynman suggested, but we conjugated them together to form entirely new mathematical operator, eliminating arbitrary performance of the coupling constant: the evolution of an event will follow to the discrete dynamic transformations of local function **(Δf)/f_1** such as **(f_2/f_{1}-1)**. This operator has important features because the field theories involve the differences of the parameters, but not input values itself, such as electricity theory involves the difference of voltages but not voltages themselves.

This formulation does not lead to the confusing approach; it is like frequency of repeating of some action in relation to the previous ground instant, giving the product of time phase **Δt/t_1**. This function is the frequency domain product, which is different from the classic physic's frequency.

Now we have to apply this concept for space and time ingredients of the space-time unit. The mathematical unit "time phase" as a frequency domain describes the displacement from a specified reference point at the **t=t_1**. The time and space phases have a commutation through product of certain position in space within some interval of time **(S_1Δt)** and with interval of space at a moment of time (t_1ΔS).

It is easy to show that the time domain in the form of time phase is equivalent and commutative to the space phase of an event and describes the relation of passage of time (**Δt**) to some local instant state or an event local boundary, such as **Δt/t_i** where **t_i** is the local time boundary of an event. Similarly, we can present the space phase with the same way **ΔS/S_i**, describing the change of space in relation to some starting local boundary of the event. On this basis, the event dynamics can be described by the shift in phase within space-time frame giving the event history equivalently in the form of direction of time or displacement of space phase. Therefore, the commuted phases may be equivalent through commutation of variables, mapped within their boundaries.

The new operator in the form (**Δf**)/**f_1** describes change (vibration) of function around its origin with certain rhythm in the form of fluctuation density to repeat its origin. Similarly, the operator ΔS/S_1 describes fluctuation of change of space energy density around space origin while operator Δt/t_1 describes fluctuation time of change around instant origin in the form of frequency.

The initial values of space and time in the form of **S_1** and **t_1** are the local dynamic boundary states: if the event within some region has ended, **t_1** transforms to the end t_2, but if the event is continued, **t_2** became a new initial state. On this basis, the local boundary is the mixture of past and future, which may have uniform and non-uniform states. Therefore, without correlation of initial and end states the action is not conserved. The non-unitary time domain **Δt/t_1** describes discrete time integer numbers of the non-continuum event. In the same way the ΔS/S_1 describes the non-continuum discrete changes of space phase with dynamic (i=1, 2, 3) boundary conditions.

The space phase in the form of integer multiplication ΔS/S_i involves the numbers of the "pieces" (grains) located in the space medium, while **Δt/t_i** describes the duration in relation to the instants corresponding to the change of position of these "pieces". It is easy to see that relation of the space and time intervals to the dynamical local state presents the transformation of continuum intervals to the discrete integers of subintervals at points $\mathbf{S}_i$ and $\mathbf{t}_i$ forming definite canonical variables of the differentiation.

This approach leads to the binding of space and time phases to each other. The relation of space and time intervals to each other usually forms the classic velocity vector $\Delta\mathbf{S}/\Delta\mathbf{t}$. However, this derivative of uncertain intervals represents the singularity at boundary conditions. The infinity in this case arises due to the absence of the commutation of the change with the local boundary of the **S(t)** function.

Interaction of time $[\mathbf{t}_1, \Delta\mathbf{t}/\mathbf{t}_1, \Delta\mathbf{t}]$ and space phase $[\mathbf{S}_1, \Delta\mathbf{S}/\mathbf{S}_1, \Delta\mathbf{S}]$ matrices leads to the formation of space-time field. Field as the correlation of the end of the phase with the initial boundary eliminates the infinity due to the conjugated displacements in space $\mathbf{t}_1\Delta\mathbf{S}$ and time $\mathbf{S}_1\Delta\mathbf{t}$. On this basis an event can be described through change of the action energy applied to the space-time field instead change of position or coordinates of the system. The $\Delta\mathbf{S}/\mathbf{S}_1$ shows distribution of energy in space: when $\mathbf{S}_1$ is small the space gets the same property within all regions of the field while at high scale, the space field is non-invariant.

By relativity, the space-time frame is curved due to the presence of energy and mass in it. But by our concept space-time field is the resulting inner product of action of the energetic field $[\mathbf{E}_{act}, \Delta E/E_s, E_s]$, formed from exchange interaction of action and event's energies:

In accordance with the Lagrangian mechanics, "an object subjected to external influence will choose a path which makes the action minimum". By our concept, this phenomenon is due to the discrete exchange of the action with the response force of the system's field, which forms the action – response non-invariance parity of the interaction. The effect of action of the applied force to the space-time field of a body can be formulated in the form of exchange of energetic fields-$(\mathbf{E}_{act} - E_s)/E_s$:

$$\frac{\frac{\Delta S}{S_1}}{\frac{\Delta t}{t_1}} = \left(\frac{E_{act}}{E_s} - \frac{E_s}{E_s}\right) = \frac{(E_{act} - E_s)}{E_s} \tag{5}$$

$$\frac{\Delta S}{\Delta t} = \frac{S_1}{t_1} \cdot \frac{(E_{act} - E_s)}{E_s} \tag{6}$$

where, $\mathbf{S}_1$ and $\mathbf{t}_1$ are the space and time variables corresponding to the dynamic local boundary, $\mathbf{E}_{act}$ and $\mathbf{E}_s$ are the energies of action and under action systems of interaction at conditions corresponding to the local boundaries of $\mathbf{S}_1$ and $\mathbf{t}_1$. The interval of time $\Delta\mathbf{t}$ describes the duration of coupling of the event with the action, while $\mathbf{t}_1$ presents the dynamic time instant or time boundary of correlation with the action. In the same way, the interval of space ΔS describes the expansion, while $\mathbf{S}_1$ presents the local boundary of space of an event at the instant. The quantity of energy, available for change (scattered energy) and the quantity, determining the response (consumed energy-momentum) have different signs (5) therefore, leads to the generation of direction and causality between local past and local future.

The equation (6), which we generated has based on the space-time consistents of energy. Model (6) although is very simple description of event dynamics but it allows to give a new look to the interaction, localized within triangle matrix of space-time-energy boundaries. In accordance with the model (6), the change is the "tree body action" of the space-time-energy operators: space derivative of energy produces momentum, time derivative conservation of identity is the energy and the product of energy derivative in conjugated space-time phases is an event. The product of energy–time multiplication is the identity of an event, localized in the observed space.

As can be seen, the change of a system's state is the result of interaction of two space-time frames, which presents the change of one field in relation to the action "field". Model (6) treats the matter field with space phase and antimatter with time phase. The time phase describes the change of event instants over time duration, but frequency domain shows how much of the time instants lie within each given frequency.

Model (6) may be re-written in the form:

$$\frac{\Delta S}{\Delta t} = \frac{S_1}{t_1} \cdot \left(\frac{E_{act}}{E_s} - 1 \right) \tag{7}$$

$$\lambda = \left(\frac{E_{act}}{E_s} - 1 \right) \tag{8}$$

In the classic formulation the change of the space-time frame **ΔS/Δt** cannot be the precise Eigen state of the dynamical event but in the model (6) it gives the precise Eigen state of identity due to its conjugation with the Eigenvector of dynamical local position $\mathbf{S_1/\ t_1}$ and with the exchange energy of the space-time variables (E_{act} / $E_{s\text{-}}1$). The reflection function ($\mathbf{E_{act}}$ / $\mathbf{E_{s\text{-}}1}$) became the Eigen value of the ΔS/Δt, which covers all the values of correlation of space-time variables. The $\mathbf{E_{act}/E_s}$ describes the density of energy, distributed in time phase as an action of force, while $\mathbf{E_s}$ has a relation to the density of energy distributed in space phase in the form of matter. Therefore, **ΔS/Δt**, conjugated with the dynamical local position $\mathbf{S_1/\ t_1}$ in the form of speed, became the function of state holding conservation of energy within space-time phases. On this basis, the model (6) became the equation of state. The equation of state (6) in its basic form (7) gives the numbers due to the resulting non-unity of the parameters.

Model (4) shows that particles do not have the fixed position during action and measurement, but they possess the change of space phase by coupling with the action energy. In this case, the measurement of the position is the measurement of the velocity and impossibility of the measurements of these parameters in different order is not the subject of uncertainty in nature.

The minimum portion of energy, which by quantum physics called "quanta", is the elementary space-time "field". The energy in similar way also is "quantized" within space-time field. The total energy is conserved discretely; it comes in discrete amounts, localized within space-time phases. The energy is the inner product of coupling of space and time fields and exists in the form of resulting Eigen value.

Model (6) shows that the space-time phases in the form of two different inner products of action –response parity display the frequency of energy consumption/restoration cycles. When there is no energy ($\mathbf{E_{act}}$=0) the coupling of virtual space and time phases generates energy, then the generated energy produces the non-virtual space-time frame of identity which conserves the produced energy within momentum. On this basis, time is the product and measure of the consumption of energy. The different frequency of energy consumption for different events leads to the separation of the non-continuum periodic events by real numbers forming the "Time phenomenon". Time takes its origin only from discrete energy conservation cycle and due to the relation to the initial state, cannot be described only by intervals. Time is the "proper time" only in connection with the Eigen value (proper value). The notion that" time slows down by distance" is the correlation of variables and this correlation is the result of discrete energy conservation.

The interaction of action and response is affected by the initial action in the form of relations of the action $(\mathbf{E_{act}} - \mathbf{E_s})/ E_{act}$ and response $\mathbf{E_s}/ \mathbf{E_{act}}$ strengths, but the outcome is determined by the advanced response given by the mathematical structure of Eigen value, which regulates energy distribution within space-time frame.

In accordance with the model (6), time appears as the personal product of exchange interaction of action-response parity and is the quantity, which holds discrete energy conservation within this parity. Therefore, relation of time to the external system of reference having no energetic parity with the system does not make sense for description of the event dynamics. On this basis, if an observer does not apply the force for interaction with an event, it became the local "photographic plate" of an event. In accordance with the model (6), the instant of time is the dynamic phenomenon and when energy is applied it produces the response in the form of inertia to regulate the continuity of the event.

In accordance with our concept, the description of a "change" in magnitude and in direction is possible only through correlation of space-time canonical variables regulated by Eigen value $(\mathbf{E_{act}} /\mathbf{E_{s-}1})$.

The classic mathematical tool of differentiation does not describe the boundary of function and the non-boundary concept of time presently is the common acceptable concept of physics. The differential operator in the absence of boundary function leads to the approximation of the action conservation due to the lost of the original function during production of the outcome of the operation.

In accordance with the model (6), when the velocity $\mathbf{\Delta S/\Delta t}$ commutes with the dynamical boundary of space-time it became the non-relativistic operator. The concept of velocity is to be tied to the space-time boundary while description of velocity without boundary leads to the "problem of different observers ". The boundary-mapped space-time eliminates the uncertainty in the continuity and the singularity of boundary conditions. This is the mathematical background of our model based on the new commutation concept.

The conditions $\mathbf{\Delta S /\Delta t=S_1/t_1}$, $\lambda_{=}1$ $(\mathbf{E_{act}=2E_s})$ are similar to the mathematical principle of involution. The inversion of the action by the same coordinate line does not produce the same action. That is why action-response parity should involve one more dynamic intermediate state, which leads to the three body interactions. The additional intermediate is the "second material observer" of (S_2, $2S_1$) interaction. On this basis the "color based strong interactions" is due to the "three sigma" color phases (colors), correlation of which in **1:2** coupling creates formation of three body interactions: the energy of two merged colors is balanced with the third color with formation of "color based elementary space-time unit". The local state with the dynamic conjugation of two merged colors with the third color leaves Eigen value the same, leading to the conservation of energy and formation of constant interactions. In accordance with our concept, virtual particle appears when the duration of change is equal to the instant of generation.

By prediction of model (6), the action of a force is to change the state of an event, but response of system (negative sign) is appeared to make the action minimum and maintain the initial state of the event.

The action of the force gradually became minimum that is why the force to be needed to maintain the initial state of action. The inversion of an event in non-virtual space-time frame in spatial and time reversal manner cannot eliminate the asymmetry of the action-effect parity. The difference of action-effect quantities determines positive time direction and magnitude of a motion (momentum).

The action-response parity is the interaction of the two fields, which "shakes" each other with the selection of the direction to move: the Eigen value of the space-time structure determining momentum of the system became the degree of freedom of the resulting Eigenvector field.

By Feynmann analysis, Dirac showed that in quantum mechanics there is important quantity, similar to the differential equation, which carries the wave function from one time **t** to another, such as **t+ε, t+2ε, t+3ε**. [7]. Feynman developed Dirac analysis and succeeded in representing quantum mechanics directly by the Lagrangian action.

Giving characterization of his research Feynman showed "that the important issue in his development of the space-time views of quantum electrodynamics is that he connected the Lagrangian with the quantum mechanics. But Lagrangian for strong interactions still needs renormalization". The problem of Feynman's Lagrangian is the non-conservation of the action. Our concept of gradually advanced response is opposite to the Feynman action and Weyl concepts where has been used the advanced action wave which violates principles of causality. The advanced response as the resulting quantity is the "hidden correlation" of space-time variables to realize discrete energy conservation.

In accordance with the model (6), space-time without coupling with the discrete energy, conservation law can give only uncertain position. This feature of space-time explains why string theory suggests additional dimensions to describe a position. The extra dimension in reality is the correlation of space-time variables with the coupling energy, which through coupling with the Eigen value rotates the space-time vectors in the form of curled up dimensions. Particle without coupling is not observable and has negative existence.

Model (6) connects position-momentum and time-energy relation and shows that these relations within space-time boundary frame cannot be subject of uncertainty because position as a spatial variable does not have existence, independent of time. Model (6) involves the commutation of first order derivatives of space and time variables from each other.

4.2. The action-effect parity of the "Change" phenomenon

By prediction of the model (6), the action of a force is to change the state of an event, but response of system (negative sign) is appeared to make the action minimum and maintain the initial state of the event. From model (6) also follows that the system, applying the energy (E_{act}) is the origin of the causal effect.

The action of the force gradually became minimum that is why the force to be needed to maintain the initial state of action. The inversion of an event in non-virtual space-time frame in spatial and time reversal manner cannot eliminate the asymmetry of the action-effect parity. The difference of action-effect quantities determines positive time direction and magnitude of a motion (momentum).

The action-response parity is the interaction of the two fields, which "shakes" each other with the selection of the direction to move: the Eigen value of the space-time structure determining momentum of the system became the degree of freedom of the resulting Eigenvector field.

With the discrete energy conservation, the forward action decreases by one while backward response increases by one (7). The advanced response force regulates conservation of energy, mapped within space-time frame. If the response (axial vector) does not change it can not limit the action. This is the non-Lagrangian least action, produced as the resulting exchange quantity.

In accordance with the model (6), the space-time frame without coupling with the discrete energy conservation can give only uncertain position. This feature of space-time explains why string theory suggests additional dimensions to describe position. The extra dimension in reality is the correlation of space-time variables with the coupling energy, which through coupling with the Eigen value rotates the space-time vectors in the form of curled up dimensions. Particle without coupling is not observable and has negative existence (6).

The Heisenberg's uncertainty and relativistic reference frame will have the same nature, if relativity's observer has an energetic interaction with an event. The observers of both theories during the measurement will have variable action (variation of consumed energy), therefore Einstein's and Heisenberg's observer's measurements are not invariant and affected by an "external force of a body under measurement". On this basis, it is obvious that the Heisenberg's and Einstein's observers cannot measure the fixed static position.

The Heisenberg's discovery on non-commutation between velocity and position is due to the problem of classic differentiation, which produces non-conservation of local velocity. Position is the integral product while the velocity is the differential outcome therefore these parameters are not commutative and are origin of the non-invariance of the action-response parity (λ). The action-response parity generates discrete energy distribution, which is only the way to eliminate infinity from space-time frame.

In the case of Heisenberg's thought experiment the measurement is an application of energy $\mathbf{E}_{act}$ to change the position for getting the information. In accordance with the model (6), the action energy portion cannot be less than energy (E_s) of the elementary space-time unit of the light. The space dimension in this case is related to the distance between the wave crests of space and time phases of the light photon. The difference of the phases cannot be smaller than wavelength of the light waves. This is the limit of phase difference of classic space-time frame, which has been called Planck's scale.

On this basis, we can unify Einstein's observers and Heisenberg's measurement. If the observers do not have energetic correlation with the event, they are distant "photographic plates" but if correlate with the event with the same Eigen value, they are invariant observers. Therefore, the Eigen value (the numbers of the scattered energy portions) "is the reference frame which determines the invariance": time is relative to an observer if the observer affects the Eigen value of an event. The Eigen value determines the intrinsic property of a system in the form of identity.

This principle of simultaneity explains Copenhagen interpretations, which states that the outcome of an experiment is only revealed when the quantum system interacts with a macroscopic apparatus of measurement resulting only one outcome.

The measurement changes the energy $\mathbf{E}_{act,}$ applied to the system and correlated with the space-time framework (6). In accordance with the model (6), you cannot measure the parameters at their fixed states, which have no independent existence. The entanglement concept of quantum mechanics may be explained also based on these principles. You cannot measure one parameter determined by space-time, fixing another one because there is no independent existence of the parameters, correlated within space-time boundary. In the absence of measurement, the system has its own deterministic space-time frame, regardless how "quantum physic's and relativity observers will observe an event on their photographic plate".

By quantum mechanics, the action is "quantized", but by our concept, the quantization is the energetic discreteness of the classic action and existence of reality by action-response parity. Conservation laws, mapped within Lagrangian framework of classic mechanics results only approximate conservation due to the non-invariance of the action. The invariance requires constant energy supply, which is not possible in continuous mode therefore the underlying mechanism of reality is the non-linearity of the physical events, realized through non-invariance action-response parity and discrete energy conservation.

5. The Non-Noether's concept of symmetry and energy conservation law

In accordance with the model (6), the classic principle "the total energy of a dynamical system involving kinetic and potential energies is conserved" can be the true concept only if the system's dynamical coupling with the action is conserved. The statement that "the energy is not created and not destroyed, but transforms from one form to another in self-sufficient system" should lead to the non-invariant energy transformation, disappearance of action and momentum. By this principle the action, kinetic and potential energies individually are not conserved and even cannot be conserved.

In our concept, we replaced Hamiltonian static sum of energy by the dynamic scattering energy where total energy, distributed within two systems, presents interaction term. The interaction term of Schrödinger equation is linear while the Eigen value of model (6), correlated with the change of local dynamic space-time boundary, determines the proper states instead of Schrödinger's probability.

Model (6) shows that the problem of Hamiltonian in canonical coordinate's transformation is directly related to the principles of energy-momentum conservation. Hamiltonian operator of total energy involves independence existence of kinetic and potential energies but in discrete energy conservation, concept energy has no independent existence. The expression of resulting energy of the system through simple gradient of kinetic and potential energies may produce only continuous displacement therefore leads to the observed problem of Hamiltonian conservation. Hamiltonian sum of energy describes the state of a system as an independence existence but in accordance with the discrete energy conservation concept, the identity cannot be described by its own existence.

The question is why we need discrete energy conservation law is very important which determines all the features of the new physics. The static continuous energy conservation described by Noether's theorem does not limit the boundary of the conserved quantity, therefore leads to the singularity.

By complementary principle of quantum mechanics, "the classical concepts such as space time location and energy-momentum can not be combined into a single picture. One classical concept excludes the simultaneous application of other classic concept. The uncertainty suggests that this reciprocal limitation is due to the uncertainty and uncontrollable exchange of momentum of a particle with the space-time frame of the object where a particle is located".

In accordance with the discrete energy conservation concept, momentum in the form of two reciprocal parameters appears for generation of the space phase of energy conservation. Conservation of energy in the space phase generates a "mass of space", as transformation of energy to mass, to generate the opposite time phase of energy conservation.

In accordance with the model (6), energy as an identity can be conserved only discretely and discreteness is realized with the alternation of the two opposite appearance –disappearance phases: energy disappears in space phase and appears in time phase. When energy is conserved in space phase, it leads to the appearance of mass. Therefore, the "mass" phenomenon is the property of the energy conservation but not the property of a body "affected by a force".

In accordance with the quantum mechanics, forces are manifestations of exchange of discrete amounts of energy. Without locality in space, exchange of energy quanta is not possible. That is why quantum physics concept that "any field of force manifests itself in the form of discrete particles" may be realized only in the presence of space medium, which display discrete existence.

From model (6), we can get Newton's first principle: Eact=0 describes the body, which is not affected by any force and localized in the inertial frame. This is the initial state where events move with the uniform motion. Therefore, E_{act} is the generator of space-time frame, which appears discretely. The state of appearance of E_{act} is the Planck scale where difference between space and time disappears. Energy is the generator of space-time and energy quanta itself is the space-time cell.

Space-time unit cannot exist alone and has to be interacted with the event where s_1/t_1 is the elementary cell of the space-time frame, $(\mathbf{E_{act} - E_s)/E_s}$) is the event. If there is no action energy Eact=0, everything is going to the initial cell s_1/t_1. In this case, the initial state and change have different sign and initial state became resources of inertia. In the absence of the action energy, the structure of the space-time changes: correlation of space and time variables t_1 $\Delta S=S_1\Delta t$ disappears which lead to the separation of t_1 from Δs. Change of space-time frame is associated with the change of energy-matter structure of an event. This leads to the separation of electron-positron and neutrino-antineutrino pairs and formation of e-/e+and v-/v pairs. The e-/e+and v-/v pairs form the pre-existing form of matter. The matter is composed from a fundamental cell of space-time frame, presented in the form of minimum S_1/t_1 cell.

6. Analysis of quantum mechanics based on discrete conservation of energy

By quantum mechanics, the vacuum energy is to be the virtual particles, which as vacuum fluctuations are created out of the vacuum. This concept does not explain what the nature of lowest vacuum energy is. The important question is why during removal of matter from the vacuum it does not reverses the energy back to the background state.

Quantum mechanics describes energy of empty vacuum in the form of virtual particles giving little push to the start of the universe, which can continue acceleration. The problem is that the calculations of quantum field theories predict 10^{120} times more quantum vacuum energy than that of any possible value.

By quantum mechanics, vacuum cannot have zero energy because of uncertainty principle, which can be violated: the zero energy value is certain. Uncertainty does not describe how the lower energy state may have high zero point energy state.

In accordance with our concept, the zero point energy means that the total energy of space-time is zero (Eact=0) but this energy is accumulated within asymmetric space-time boundaries. The total energy is distributed within different phases, which are in asymmetric state of opposite boundaries.Vacuum and black hole are two asymmetric boundaries of space-time variables. The theories of quantum physics suggest that due to the uncertainty principle, quantum field cannot have zero value, therefore became the origin of vacuum fluctuations. In accordance with our concept the vacuum and "quantum fluctuations" is due to the discrete conservation of energy within asymmetric space-time phases. At small scale of space, the frequency of fluctuations is generated by time phase and is high while at high space scale the giant size of fluctuations is created by the energy consumed in space phase.

In accordance with our model (6), creation of a particle is the result of discrete conservation of energy, which appears within space and time "fields". Conservation of energy is not static, therefore mass also is not static and changes with the expansion of space. This feature of mass is the necessary factor for conservation of energy; with the constant mass, energy cannot be conserved. On this basis, vacuum is the asymmetry of space-time boundaries, therefore light cannot exist in vacuum with the constant static speed. Vacuum appears when the energy of the background state is removed for expansion of space. When energy is completely consumed in space, time returns it back to the initial state by negative gravitational force. That is why we replaced the spatial and time coordinates by space and time phases having performance of the energetic fields.

By quantum field theory, vacuum has properties as a particle and these properties cancel out on average, leaving the vacuum empty. By our concept vacuum is the asymmetric boundary of space-time where space identity is very small which is connected with the high amount of time particles –antiparticles.

The classic physics relates local symmetry to the Lorentz invariant quantity, which is connected with the helicity phenomenon, describing the projection of a spin of a particle in the direction of momentum. In the case of discrete energy concept, local invariant inversions are not allowed.

Discrete conservation of energy in space –time phases cannot be realized without boundary of theses phases which constrains the expansion of space through left handed neutrino (involved in Es) and right handed anti-neutrinos (involved to E_{act}). That is why the right handed neutrino and left handed antineutrino never was observed.

In accordance with the Newton's physics, mass is the inertial rest energy of a particle and the measure of the resistant to the applied force (a=F/m). By special relativity, a massless particle cannot exist at rest it must always move at speed of light. Quantum mechanics suggests that a "massive fermions should have both right and left hand states because field operators that yield a non zero mass for fermions are bilinear products of fields that flip the particles handedness".

In accordance with the discrete energy conservation, the quantum mechanic's mass operator, which annihilates left handed neutrino and creates right handed antineutrino can not be described by the sum of doublet due to the continuous nature of this interaction. Therefore, the mass term that changes a particle into antiparticle, which in quantum mechanics called as Majorana mass term, has to be described by discrete term, such as (E_{act}/E_s-1). The sign in this formula changes (such as quantum mechanic's fermions' number, which changes from n=-1 to n=+1) right-handed particles to the left handed.

The uncertainty of a position and the future motion of a particle can be accepted classically obvious, because position is not a static quantity. It is known that measurement of the particle position involves the scattering of light, which can give only probabilistic exchange of energy and cannot be described by known classic physics. Quantum mechanics explains, "If scattering energy is not uniform the measurement devices has no possibility to measure the position and momentum".

The problem is we cannot measure position with instant of time only and measurement should involve instant of time and its change in the form of time phase $\Delta t/t_1$. In accordance with the model (6), the change of time in relation to the instant may describe the change of a position and velocity. The change of time phase has no independent existence and its changes in correlation with the change of space phase. That is why measurement of the space and time as the independent coordinates leads to the uncertainty. Therefore, the mathematics of the description of the position and velocity should involve them not as an independent coordinates but correlated phases.

Therefore, the statement that classic concept does not fit at the quantum level is not true. If time will involve two parameters as an instant and duration, the ratio of this parameters may describe any level of an event: at small scale the duration is small and instant has high frequency while at high scale instant is small duration of an even is long. At "quantum level", the high frequency instants are not quantum jumps and they are discrete phases, appearing in space phase and disappearing in time phases. The alternation of these phases and exchange of energy between these phases can not be described by probabilistic energy distribution, because discrete exchange of energy is very causal deterministic process.

Heisenberg uncertainty implies that "any two variables that do not commute can not be measured simultaneously". It is possible to give classic explanation to this phenomenon. Two variables that have no independent existence cannot be measured simultaneously. The position and velocity have no independent existence similarly as change of energy is not independent of time. Therefore, uncertainty of quantum mechanics is the non-conservation of energy in the formulations of classic physics. That is why "uncertainty of non-commuted variables" is the uncertainty in conservation of energy in the farme of classic physics.

The question is how to describe displacement of space boundary within certain time interval or displacement of time boundary within certain space boundary. On this basis, the main concept of quantum mechanics is the "quantization" of space-space period.

In accordance of our concept, quantization of space-time frame is the classic phenomenon and it is related to the discrete conservation of energy, mapped within two opposite phases. "Quantization" of space-time frame requires correlation of space-time variables, appearing as a field, realizing the dynamic state of energy conservation. Model (6) shows that connection of space and time together in one unit, is the quantization of space-time, which puts limitation to the boundary of space-time. However, for connection of space and time in space-time unit, free coordinates cannot present these variables.

The non-independent change of space-time variables leads to the displacement of space and time phases by the alternation, which generates a wave function carrying the portions of conserved energy in different phases.

In the case of discrete conservation of energy, the particle-antiparticle annihilation is not a symmetric event and cannot be described by symmetric model. Therefore, the symmetric model leads to the continuous energy conservation which results "Ultraviolent Catastrophic" phenomenon of the background energy. Due to this problem, renormalization is an open question in the quantum field theory where the total energy becomes infinite.

By quantum mechanics, the zero point energy is the "lowest" quantized energy level of a quantum mechanical system. The zero point energy is "the energy which remains when all the energy is removed from a system". The common opinion is that the origin of the zero point energy is the uncertainty principle. By quantum physics a quantum fluctuations is the temporary change for energy in a point in space, arising from Heisenberg's uncertainty principle, "such as temporary change of amount of energy in a point in space without certain time". The uncertainty accepts that conservation of energy can be violated but for small times. In accordance with the energy-time uncertainty, with the decrease of time, violation of energy conservation increases. Therefore, the uncertainty principle allows existence of a virtual particle with the barrowed energy, but the sources of the borrowed energy are not clear.

Quantum mechanics suggests that pair production appears as a pop into existence and then annihilation each other. On this basis there is suggestion that [8] energy in classic physics is conserved properly but in the quantum micro world energy can appear and disappear in a spontaneous fashion. The uncertainty principle implies that particles came into existence for short periods even when there is not enough energy to create them. They borrow energy for a short time and then they return and disappear again. In vacuum, pair of virtual particles is constantly being created and destroyed. The energy of matter is positive which appears spontaneously out of empty space while the energy of anti matter is negative. The matter made of positive energy and matter particles has attracting gravitational energy, which is negative. The total energy is zero. By quantum mechanics, fluctuations are random and have no causal nature.

Quantum theory of the vacuum suggests that the zero state of vacuum energy is negative. Based on expansion of the space it concludes that vacuum ground state has non-zero energy. The vacuum energy, described by quantum field theory without renormalization is mathematically infinite.

It can be thought that discrete energy conservation is obvious concept due to the discrete dynamics of quantum mechanics and discrete energy radiation. But the question is how to describe the discrete energy conservation in mathematical formulation. In accordance with our model (6), without correlation of space-time variables, it is impossible to describe discrete conservation of energy. With the frame of discrete energy conservation, it is impossible to describe spatial or temporal inversion without boundary space-time framework. In accordance with the discrete energy conservation, the space-time dynamics appears as the particle-antiparticle pair, which carries a conserved quantity of energy within these phases.

Quantum mechanic's theories suggest that zero point energy has positive sign for bosons and negative for fermions, which cancel each other for perfect symmetry. The model (6) describes the energy of bosons and fermions with the similar way but shows that they cancel each other at Eact=2Es.

For uniform speed of light, the change of space (ΔS) and time (Δt) have to be correlated, generating invariance of these intervals from the boundary of space-time frame and uniform change of these variables. The energy of light at uniform speed should follow the Noether's theorem of symmetry. It is known that Maxwell equation is not independent from the property of distance and the speed of light is a function of properties of space. Model (6) shows that property of space (ΔS) and time (Δt) intervals changes with the change of boundary of these variables.

In accordance with the quantum mechanics when matter and antimatter collide, they annihilate in a flash of energy. In accordance with our model, the energy, produced during interaction of asymmetric space-time phases has to be conserved in space phase to form "particle" which disappears in the opposite time phase, forming "virtual particles". The observance of these particles depends from frequency of the discrete energy conservation.

7. The singularity problem of relativity and uncertainty of the quantum measurements

Due to the lost of the initial boundary frame, all space-time models break down at the singularity. To eliminate the singularity, the space-time should have internal constraint holding the dynamics at boundary, because the initial position is not an independent free coordinate. The initial state of minimum space-time identity is the internal constraint of the space-time, which determines how to go from contraction phase to the expansion phase without singularity. General relativity has fused discrete space-time frame to the continuous "geometrical manifold" which with the decrease of space has to move to the infinite space-time curvature. The space-time frame by this concept has no internal constraint therefore has to move to the singularity and the initial state of universe should start from a single point having infinite density. This is the unproved conjecture of general relativity.

At continuous energy conservation, an event with any size has a trend to move to the equilibrium: to remove a system from the equilibrium one additional parameter is needed which can be the application of an external energy. With the continuous energy conservation law a system, consuming any amount of external energy has to move to the equilibrium of "infinite black hole".

The asymmetry of space-time boundaries is the fundamental basis for discrete existence of universe. In some theories, dynamical space-time has been discussed but there was not shown that this dynamics is due to the asymmetry of boundaries. The discrete coupling constant $\mathbf{E}_{act}$ (magnitude is finite) is the boundary constraint of space-time which at $\mathbf{E}_{act}=0$ is different from Einstein's $\mathbf{R}_{ab}=0$ constraint of empty space. The discrete conservation of matter and energy within discrete space-time frame gives a non-geometrical status to the space –time phenomenon.

In accordance with the model (6), inverse transformation appears when all the energy of space-time is consumed by space ("black hole phase") and all the events have the same dynamic characteristics of "free fall "to the state of energy generation".

8. Particle physics based on discrete energy conservation

Presently why the particles acquire the mass is not known. The mass –energy relation which first time has been introduced by Einstein in the form **m=E/c²** does not describe mechanism of mass generation. The problem of this formulation is invariance of mass-energy relation, fixed by speed of light.

As follows from the model (6), conservation of mass cannot be separated from the discrete energy/momentum conservation. Therefore, in discrete conservation of energy there is no invariant rest mass, mapped within discrete space-time frame. The problem of mass appears with the Newton's force. Newtonian law does not explain the origin of force that is why the nature of mass remains unanswered. Newton's invariant mass during change of velocity leads to the approximate energy conservation. If force is needed to change the velocity, then Newtonian inertia appears to describe the magnitude of change in the form of mass. If the action of force, as is shown in (6), is needed to hold the initial state of a body then it is needed to hold the existence of matter in space-time frame in the form of mass. The problem of Newtonian law is that mass, which is invariant during velocity differentiation but affects to the acceleration.

In accordance with the model (6), the entity called mass is the resulting quantity, generated from discrete energy conservation. This feature seems makes it as the "electromagnetic mass". With cyclic discrete energy conservation there is no any preferred reference frame, which could display inertial or relativistic mass. The non-virtual matter particle is created simultaneously with the generation of non-virtual space-time frame: mass is the part of energy to be conserved in space phase: the dense space phase, formed through change of frequency of correlation of space-time variables leads to the generation of different masses responsible for different particles. The density of space phase is determined by the Eigen value (6), therefore mass is not constant, it changes with the change of frequency. Due to this phenomenon electron has no constant mass, it transforms to other families as it moves through space.

Discrete energy conservation mapped within discrete space-time phases has two consecutive steps: a) re-alignment of minimum space phase, "b) accumulation of mass in space phase.

Discrete energy conservation eliminates the infinity of the energy, produced during contraction of space phase to minimum size (6) through re-alignment and coupling with the time phase. The time reversal asymmetry of space-time boundaries arises from discrete energy conservation where forward and backward directions have different initial energetic conditions. Fraction of time is needed to complete a full cycle when the minimum portion of the space phase passes through inversion frame from negative to the positive direction. The time reversal transformation of matter/antimatter particles can be described as follows:

$$\gamma/\gamma = -\left(e^{+}/e^{-} + \nu_e\ /\nu_e^{-}\right) \tag{9}$$

Formation of two pairs of particles from one pair (9) is due to the" three body" identity conservation, described by condition (9).The equation (9) describes total spin symmetry of matter-antimatter interactions. The spin of right-handed photons is equal to the sum of the two half spins of left-handed particles which hold conservation of energy, charge and momentum. The super partner of gamma rays is two pair of (four) fermions (9) which generates re-alignment of the particles in space-time reversal cycles. The ingredients of the equation (9) form the triplet spectrum generating re-alignment of light and dark photons on intermediate $\mathbf{e^+/e^-}$frame.

As can be seen, without composite photon it is impossible to describe generation of mass. When neutrinos interact with the charged particles, fermions are generated but when they interact with each other "gamma ray bosons" are formed. The three families of particles participate in generations of right side" pseudo-bosons" and left side composite photons.

The mechanism of equation (9) is in agreement with the results of Stefano Profumo[9], who found that electron, reacting with the surrounding dark matter could fuse into a heavy version of electron and when it returns to the initial state radiates gamma rays.

It is necessary to note that the quantum mechanics concept of transformation between photons and $\mathbf{e^+/e^-}$does not preserve energy conservation due to the non-locality of energy transformation. Light quanta do not have independent existence and cannot hold fixed amount of energy without space-time structure.

In accordance with the scheme (9), discrete inversion between light photons and neutrinos (dark photons) is to be realized through electron/positron pair playing a role of "intermediate mixing space-time frame of matter-antimatter pairs" which leads to the generation of charges. On this basis, it is clear why matter-carrying ingredient of Eigen value is negatively charged. In accordance with the scheme (9), merging and co-vibration of light and dark (neutrino) photons in mixed frame produces charges $\mathbf{e^-/e^+}$. On this basis, photons and $\mathbf{e^-/e^+}$pair show behavior in a similar fashion being origin of each other and light photons have no motion without alignment of electrons with the neutrinos. The inversion model (9) of the matter/antimatter transformation is different from Dirac vacuum, containing only electron/position "see".

The scheme (9) involving inversion operations describe charge and spin conservations. In accordance with the equation (9), electrons are the light photons, localized in the virtual space-time frame that is why all electrons have the same charge. When a particle form non-virtual space-time frame it became a different particle.

In accordance with the condition (9), photon has a virtual leptonic structure and the same time the pair of leptons $\mathbf{e^-/e^+}+\nu_e/\nu_e^-$ has a performance of virtual bosons (called Nambu Goldstone bosons). The sum of $\mathbf{e^-/e^+}+\nu_e/\nu_e^-$ (9) in the form of four fermions describes the "fermionic quanta". Due to the discrete energy conservation, the amount of "bosons", distributed in space-time condensate is not unlimited and the condensate should radiate photons, similar to the "black hole" radiation of space energy.

When $\mathbf{E_s}$ increases, it absorbs more gamma rays and leads to the formation of "black hole structure" which cannot settle to the stationary state. From model (6) follows that at $\mathbf{E_{act}}=0$ "black hole" has to radiate its energy back to the minimum space state. The radiation of black hole energy happens with transformation of $e^+/\nu_e + e^-/\nu_e^-$ to the longitudinal wave of neutral ingredients $\nu_e/\nu_e^- + e^+/e^-$.

From model (6) and scheme (9) follows, that removal of last portion of matter from vacuum needs more energy than background state. Therefore, the boundary space-time unit participates in the coordinate inversion and grows of particles, which leads to the expansion of space.

In discrete energy conservation concept the initial and end states are vector fields. Absorption of photons by $e^-/e^+ + \nu_e/\nu_e^-$ pairs leads to the transition of "bosonic" structure to the frame of $e^-/\nu_e^- + e^+/\nu_e$ virtual quarks with transformation of photons to "gluons" existing in the form of meson field (pions). We suggest that the phenomenon called "spontaneous symmetry breaking" is the change of space-time structure with transition of energy within asymmetric space-time boundaries while symmetry is the derivative of energy conservation. The consumed energy is not equivalent to the scattered energy, which leads to the asymmetric space-time boundaries. The space phase (virtual fermions) is less symmetric than time phase, therefore energy conservation has handiness in direction of energy consumption and the consumption of energy is accelerated.

The interference of space-time phases at background state (9) is not linear and coupling of the matter-anti matter particles forms intermediate loop boundary of mixed matter-anti matter frame: electron is produced from neutrino side while positron is generated from photonic side. Spin of the matter-antimatter loop to the minimum size leads to the re-alignment of the phases with generation of $e^-/\nu_e^- + e^+/\nu_e$ quark/antiguark structure. Due to the different spin for bosons and fermions, the alignment is directed to the matter side with the shift of handiness from the left to the right (consumption of energy in space phase). This is the origin why matter spin has to be twice less than bosons. Therefore, the asymmetry in spin and in the boundary of space-time variables is the origin of the matter of universe. The transformation from spin one to half spin has to lead to the energy loss but this energy is accumulated in the pion and kaon fields.

Leptons are products of virtual space-time while baryons are location of quarks in the non-virtual space-time frame, which makes radiation of matter and energy non-invariant. The quantum theories treat right and left-handed particles symmetrically which leads to the invariance of Lagrangian action. This is the problem all of physical theories. Background state (9) connects all conservation laws except asymmetry of space-time boundaries. The general principle of super symmetry and action invariance can be realized when the change of the energy of bosons, relative to its initial state, is equal to the change of the energy of under action fermions relative to the applied energy. In this case, the laws can remain the same for both systems.

Quantum mechanics considers that empty space is a dynamic medium, which is full of virtual particles where "lowest energy state of the system is not zero and an isolated particles traveling in empty space interacts with the vacuum and produces virtual particles/antiparticles pair, coming from the vacuum itself. The original particle then reappears when the particle and antiparticle meet and anhillate each other". Uncertainty principle suggests that particle/antiparticle pair live on borrowed energy within short time therefore violation of energy conservation takes place only in a short time. The question, which has to be answered here, is the nature and origin of the borrowed energy. The unanswered question of quantum physics is how to describe energy of empty space if contains any kind of energy.

In accordance with the model (6), the space-time can exist only through interaction with other space-time field. The baryonic space-time structure of matter exists in interaction with the non-baryonic matter field forming leptonic matter. Light photons move through sea of e-/v...e+/v waves with constant speed and expand this boundary through generation of new portions of dark energy. When it expands the boundary of space, (e-/v…e+/v) it leads to the formation of red shift.

The existence of neutrinos in three flavors requires the existence of light photons also in three flavors similar to the equation (9):

$$G/G= - \left(\mu^{+} / \mu^{-} + \nu_{\mu} / \nu_{\mu}^{-}\right) \tag{10}$$

$$Z/Z= - \left(\tau^{+} / \tau^{-} + \nu_{\tau} / \nu_{\tau}^{-}\right) \tag{11}$$

The **Y**, **G** and **Z** describe different flavors of photons having different frequency. At this state, the energy is not dispersive (Eigen value is minus one) and the virtual photon particles are indistinguishable.

Based on scheme (9-11) we may extend our analysis on the wave –particle duality of the light. Light is wave, but presently it is not known how it is waving. In accordance with the special relativity light travels through vacuum with constant speed but this concept does not explain what medium has a vacuum, which carries the light with the constant speed. Maxwell equations also do not involve any medium to carry light waves.

Due to the discrete energy conservation, light travels in space-time phases through polarization of wave and particle properties. In accordance with the equations (6) and (7), light does not propagate without momentum which constraints propagation of light wave in space. The space-time frame and light energy radiation are symmetric in discrete mode. In accordance with the scheme (7-9), light has virtual quark's structure and does not travel in empty space.

Connection of discrete energy conservation with the boundary mapped space-time frame explains the mystery of "increase of mass with the velocity": increase of mass with the velocity was suggested by Einstein to hold constant speed of light. Discrete energy conservation and boundary mapped space-time concept describe this problem differently: light is the discrete identity, which produces its derivative space-time frame, and any observable identity is the interaction with this frame. Therefore, light does not exist independent of space-time frame.

In accordance with the discrete energy conservation concept, it is impossible to shift space-time phases of a particle having the same spin unit of its constituents. Spin is the result of energy conservation in different phases, which leads to the generation of direction.

At discrete conservation of energy, the background state of annihilation (9) does not lead to the divergence due to the conservation of finite amount of energy within asymmetric boundary of space and time phases. The alignment of the ingredients of the scheme (9) is the polarization of neutral leptons to fermions. Coupling of γ/γ rays with the **e^-/ e^++v_e** $^-$/ν pairs leads to the alignment of scalar field to the vector bosons which conjugates left handed neutrino with the left handed electron and right handed antineutrino with the right handed positron with formation of e^-/ν $_e{}^-$ and e^+/ν quark/anti quark pairs. The alignment of three e^-/ν $_e{}^-$ and e^+/ν quark/anti quark pairs in 2:1 (λ=1) mode leads to the non-integer electric charge carrying baryon's space-time structure. The energy of gamma rays in the form of "gluons" connects quark/anti quark pairs within meson field of pions. The electric charge keeps photon in the form of gluon, when charge is annihilated gluon also disappears. At decay process (E_{act}=0), the energy of gluons is consumed for re-alignment of e^-/ν $_e{}^-$, that is why electron in the resulting process appears with less energy The continuous spectrum of beta decay is due to the absence of correlation which generates left handed helicity of neutrinos. In real world due to the discrete correlation with the applied energy, the anti neutrino has right handed helicity.

Coupling of ingredients of the equation (9) at boundary loop leads to the formation of dense space particle-top quark/anti quark pair which in accordance with the model (6) leads to the energy-matter interaction with generation of virtual ingredients of baryons (**tt-+e^-/ e^++ν $_e{}^-$/ν → bb$^-$+uu$_-$**). This is the reason why top quark's mass is very close to the energy scale of alignment (symmetry breaking). Coupling of top quark with up quark takes place at **λ=1**. In this exchange interaction, the top quark is conjugated with its weak isospin partner of bottom quark. The boundary energy of top quark does not limit frequency of decay during its coupling with the up quark. Due to this function, top quark does not form baryon's space-time frame. In accordance with the scheme (9), without boundary space the alignment γ / γ→ **W^-/W^+→ TT^-**is not possible. Generation in the mixing loop (9) gives to the electron mass more than neutrino where the absorbed neutrino became the space locality of the mass-carrying electron **e^-/ν $_e{}^-$**.

Therefore, particles, which exchange energy, are bosons while particles exchanging momentum are fermions. With other definition particles carrying energy in time phase are bosons but conserving energy in space phase are fermions. The exchange of energy does not affect the wave function (Eigen state) while exchange of momentum changes the sign of the wave function. The phenomenon, which make light to move and generate discrete radiation is the discrete energy conservation within space and time phases.

9. Generation of quarks — Why the top quark does not form baryon?

The space-time frame (6) at **$E_{act=}$**0 transforms to the virtual space-time frame which by radiation of matter contracts to the minimum space size :

$$\frac{dS}{dt} = \frac{S_1}{t_1(-E_s / E_s)} \tag{12}$$

At (**$E_{act=}$0, λ=-1**) re-alignment of matter ingredients leads to the radiation of mass, consumed in the space giving the same "Femi weak coupling constant G_v "of neutron and muon decay.

The "non-correlated" space matter decay (Kaon and other mesons) generates neutral current similar to the photons. The model (6) in this case explains why correlated electromagnetic interaction of proton with electron is stronger than the reverse weak force of neutron decay to proton and electron. The alignment of the fermionic super partners $\mathbf{e^+/\nu_e + e^-/\nu_e^-}$ to the bosonic super partners $\mathbf{\nu_e/\nu_e^- + e^+/e^-}$ leads to the formation of neutral current of decay

$$\gamma^*\left(\mathbf{Gluons}\right)\mathbf{+e^+/\nu_e + e^-/\nu_e^- =- \gamma/\gamma + (e^+/e^-) + \nu_e\ /\nu_e^-} \tag{13}$$

which correlates the helicity of elementary particles with the negative Eigen value.

In classic way, a matter is the quadratic product of the two fields and coupling of space and time phases gives the potential of the observable frame. This principle is similar to the quantum mechanics concept that the square modulus of wave function is associated with the probability of observing the object.

In quantum field theory, the energy of vacuum is zero, but formation of mass requires the positive sign of energy between the vacuum and the next lowest energy state.

The generation of energy as a quadratic product of coupling of opposite space-time vector fields can be described as follows

$$\left(\frac{dS}{S_1}\right)^2 = \left(-\frac{dt}{t_1}\right)^2 \tag{14}$$

The quadratic quantity of space phase describes the energy density in space while the quadratic quantity of time phase describes frequency of the change of this phase. Coupling of opposite charge carrying space-time vector fields is similar to the coupling of matter-antimatter particles (6):

$$\mathbf{(\gamma/\gamma)^2 = [-(e^+/e^- + \nu_e\ /\nu_e^-)]^2 \longleftrightarrow (\gamma^* + e^+/\nu_e\ + e^-/\nu_e^{-)2} \text{«} (t/t^- - bb^-)} \tag{15}$$

Coupling of matter-antimatter particles of the equations (9) is similar to the scheme (13). Here arises very important question why background space-time coupling should leads to the generation of top-bottom quarks. The top-bottom quarks are similar to other quarks but due to the high energy, they decay fast and do not form non-virtual space-time structure of baryons. Therefore, $\mathbf{t/t^- \text{-} bb^-}$ interactions may exist within high frequency annihilations generating in time phase $\mathbf{t/t^-}$ particles and in space phase $\mathbf{bb^-}$ particles with the realization of $\mathbf{t/t^- \text{-} bb^-}$ meson structure. But energy-momentum frame in simple background $\mathbf{t/t^- \text{-} bb^-}$ transitions is not invariant, therefore coupling of this pair leads to the formation of another pair of quarks, for example $\mathbf{uu^- \text{-} dd^-}$ interactions with exchange of vector bosons. The magnitude of the background energy (9) is fixed within boundary of space-time that is why top quark mass does not move to the infinity.

In accordance with the energy conservation (9), **d-u** transformation of quarks cannot be realized by simple emission of the $\mathbf{W}^-$ bosons because in this case the energy is not conserved. Therefore, energy and identity conservations in quarks transformations have to be realized through coupling of **t-b** transformations with one side with **u-d,** from other side with **s-c** quarks where the total energy and momentum are discretely conserved:

$$\mathbf{tt^- + uu^- \rightarrow dd^- + bb^-} \quad (16)$$

$$\mathbf{ss^- + bb^- \rightarrow tt^- + cc^-} \quad (17)$$

$$\mathbf{dd^- + cc^- \rightarrow uu^- + ss^-} \quad (18)$$

In accordance with the equation (9), the identity of a particle can be determined with the charged particles but electromagnetism is carried by neutral photons. On this basis the particle performance of baryon in space-time frame, is carried by charged particle that is why the electromagnetic force cannot hold alone the identity conservation of baryons.

During energy transfer (16-18), the bottom quark travels as a part of jet stream but then returns back to generate cycle as "the Feynman loop". In accordance with the conditions (16-18), conservation of causality of action is realized with correlation of end and initial states through **[T-B]** bi-mesons [($\mathbf{ss^-}$-$\mathbf{bb^-}$) – ($\mathbf{tt^-}$-$\mathbf{uu^-}$)] which regulates discrete exchange of energy within nucleons and background polarization state.

Top meson has to be created from anti top and up quark while anti top meson forms from top quark and anti up quark. **B** meson simultaneously is produced from b-anti quark (**b-bar**) and **d**-quark. Its antiparticle **B-**anti meson is formed from **b**-quark and anti **d**-anti quark. These transformations take place at $(\mathbf{E_{act}}-\mathbf{E_s})/\mathbf{E_s}=1$. The discrete performance of two nuclear frames is the requirement of discrete identity conservation of nuclear within **uud-udd** and **ssc-ccs** nucleons coupled with the **t-b** transformations. The **ssc-ccs** play a role of isotopic nucleon.

The exchange of the two nucleons generates a field in direction $\pi \rightarrow$ Kaon$\rightarrow$ **B** mesons, which decays back at $\mathbf{E_{act}}=0$. This is the inverse spontaneous symmetry breaking. The stability of nuclear against to the repulsion by electromagnetic force is due to the discrete coupling of the recycled energy ($\boldsymbol{\lambda}$**=1**). This is the explanation why nuclear force is short ranged. Depending from Eigen value, meson field can be real or virtual particle. At none zero $\mathbf{E_{act}}$, pion is a particle, but at $\mathbf{E_{act}}=0$ it became field of gamma rays. At this condition, there is no difference between electromagnetic force and gravitation. The heavier the meson the shorter is time allowed for the exchange process therefore only asymmetry of space-time boundaries allows existing of heavy top mesons. The mystery of conventional physics why the pion should have mass is explained with the discrete energy conservation within matter-antimatter particles, located within space-time phases.

The constant coupling with the background energy resources leads to the internal isospin symmetry SU of baryon quarks existing in three flavor of space-time structure $\mathbf{E_{act}=2\ E_s}$. The baryon alone is not stable therefore cannot be the fundamental matter: two nucleons are coupled with the background bi-meson field **[T-B]** to form three flavor of space-time structure $\mathbf{E_{act}=2\ E_s}$. The spin of quarks within baryons is aligned to conserve the condition $\mathbf{E_{act}=2\ E_s}$ through quark/ant quark mesons. Separation of quark-anti quark pair is impossible due to the impossibility of separation of energy, distributed within space and time phases.

It is very interesting that the identical quarks do not obey the normal rules of quantum statistics as spin ½ particles: quarks should be fermions with anti symmetric wave functions but the pattern of observed baryons shows that they have symmetric wave function. In accordance with the quantum principles, quarks would radiate energy of interaction and dissipate their motion. Model (6) explains this phenomenon: the constant motion of quarks in baryons is in invariant coupling with the background state: at constant coupling with the background, state uniform transformation within **p-n** pairs gives to the quarks identity conservation with the symmetric wave function. At these conditions, the quarks have the same symmetric performance with the particles of background photons.

By quantum physics, it is hard to keep electric charge in a small pack because it repels itself. The problem of this concept is that electron /positron pair cannot be created from photons by two body reverse interactions $\mathbf{2Y=e^-/e^+}$due to the violation of energy conservation. The three-body model shows that the mixed "electron /positron bubble" is a composite frame (9) where the energy, momentum and spin are balanced properly. In the mixed frame the three components form the "bubble space-time" with generation of electron with negative sign while anti particle gets positive charge. Photon and electron ingredients of the background state (9) are the dynamic products of the geometric object – space-time vectors where generation of space $\mathbf{S_1}$ transforms antiparticle to particle while generation of time instant t_1 transforms particle to antiparticle. The three body flavor interaction $\mathbf{E_{act}=2\ E_s}$ with the discrete non-Noether's symmetry leads to the generation of ordered structure of baryons.

The momentum of an electron is generated from E_{act}/E_s interactions therefore there is no static mass and photon alone is not the electrodynamics force mediator.

10. Why the weak force is needed?

In accordance with the model (6), the phenomenon called parity does not describe inversion of a position of a static body through spatial coordinate but the parity of two dynamical space-time systems. Violation of parity is the result of non-invariance of the action-response of two systems, which generates direction of time: without local violation of parity, the global symmetry is not conserved and there cannot be energy conservation and direction of time. That is why P violation is associated with the **CP** violation. When the energy of the initial state is completely consumed in space phase ($E_{act}=0$), parity transformation changes the algebraic sign of the space coordinate to return the system back to the energy generation state.

In accordance with the gauge filed theory, the gauge bosons have a parity **P=-1,** but spin ½ particles have **P=+1**. The vector bosons and fermions to acquire masses the gauge invariance Lagrangian and new scalar boson Higgs field is predicted to exist. We suggest that with the non-invariance Lagrangian the intermediate **W** bosons participates in decay process and decay leptonically when λ=-1.

The model (6) connects strong and with weak interactions as the resulting quantity from action –response parity while Yukawa formula [10] of the strong force does not reveals the resulting quantity.Yukawa suggested that nuclear force could not be reduced to the electromagnetic interactions between charged particles. The same conclusion follows from Gell-Mann mechanism of quark interactions locating quarks in baryon octets within certain quantum numbers. The important conclusion of Gell-Mann [11] is that the strong force between quarks will degenerate if the strong force is the only force holding the quarks within nucleons.

In accordance with the model (4), one force cannot conserve the strong nuclear force and this force has to be uniformly coupled with the energy, recycled to the nucleons. In this case, the coupling constant does not diminish with the change of the space within the meson field.

By quantum field theory "when coupling constant is much smaller than one (**g < 1**) then the theory is said is weakly coupled. If the coupling constant is of order of one or larger the theory is said to be strongly coupled". The model (4) meets this statement of quantum field theory. When the Eigen value is equal to one the system is strongly coupled with the action. If $\mathbf{E_{act}}$ is smaller than $\mathbf{E_s}$ the system has a trend to undergo to the β-decay. This condition corresponds to the negative value of Eigen value. When Eigen value is plus one the particle's momentum, spin is aligned, and a particle is right handed. If the Eigen value is minus one the spin and momentum counter aligned, therefore the particle is left handed.

The important question here is why fermions are half integer particles and should obey Pauli principle. In accordance with the model (6), this question is related to the realization of constant action in three jet events and formation of space-time frame: $E_{act}=2E_s$. The three jet events are the result of space-time frame and one of the three jets has unique property under the strong interactions to realize coupling with the constant energy. Therefore, coupling holds the hadronization of quarks and without coupling hadrons' space-time frame transforms to the virtual manifold. That is why strangeness is conserved during creation but is not conserved in decay process. Based on the mathematical structure of the model (6), at constant coupling the quarks can never be liberated from the hadrons.

As follows from the equation (6), four fermion ingredients at $\mathbf{E_{act=}}0$ transforms to the neutral bosonic particle with total spin 2 which in the form of "graviton" leads to the conservation of the background energy.

11. Strong interactions and behind the standard model

Standard Model describes the origin of forces in term of local symmetries. But energetic field has to change discretely which leads to the discrete change of local space-time frame where the energetic radiation field ($E_{act}/E_{s\text{-}}1$) is associated with the asymmetric local space-time field. Model (6) shows that every local space-time frame is commuted with the energetic field, which constrains each other from the linear displacement. On this basis, the space-time frame and energy field have no independent existence and their distribution is determined by discrete energy conservation through background state. This eliminates the linear effect of the Newton's force and leads to the rotational dynamical motion (spin) of a particle around its generation frame. Model (6) shows that radiation field ($E_{act}/E_{s\text{-}}1$) and particle can be classic commutative variables which by Dirac were accepted as a commutation of quantum variables.

The space-time model based on discrete energy conservation produces symmetric and anti-symmetric dynamical functions describing the direction of decay and the nature of produced particles. In accordance with the model (6), the symmetric and anti-symmetric functions can be transformed to each other depending from energy and identity conservation laws: when the Eigen value is positive, the function leads to the hadronic transformations. When the E_{act}=0, the function (6) being anti symmetric leads to the leptonic decay.

In accordance with the scheme (9), the alignment of the photons with the ½ spin particles ($e^{+}/e^{-+}\nu_e/\nu_e^{-}$) moves the resulting energetic field in direction of space expansion with formation of hadronic ingredients. The resulting hadrons form "jet" along the direction of parent "electron particles".

The action-response parity of the model (6) shows that all the interactions (four forces) can be divided into two groups: two dimensional two body performance (matter –anti matter annihilation) with **SU (2)** symmetry group and three body interactions with commutation of **SU (2)** symmetry with **U(1)** group giving three body **2:1** confined resonance particles. The" two body" performance with **λ=-1** describes the interaction of virtual space-time annihilations (11) where the input function produces the opposite outcome. The total energy, distributed in **2:1** and **1:2** combinations of space-time waves generates 1/3 and 2/3 fractional frequency of waves which is displayed in the form of charges. Due to the operation of space-time waves within discrete symmetry, all electric charges have the symmetry in their unit.

Very important question, which may arise from three family 2:1 resonance performance, is Pauli Exclusion Principle, which states that two fermions such as quarks cannot occupy a quantum state at a given time. Model (6) shows that the existence of the two quarks with the symmetric wave function in classic way requires discrete correlation one of the quarks in proton-neutron baryons with the discrete conservation of the action.

Yukawa showed that [10] for some reason nuclear forces are saturated which lead to the appearance of different concepts for explanation of this phenomenon. Particularly Yukawa's meson theory appeared to explain this problem.

By prediction of the model (6), the symmetry between different spin particles such as bosons and fermions is possible if to accept that the force-carrying particle is not the single boson but is a complex "particle" having the sum of two half-order particles. On this basis the **SU (2)** discrete symmetry of proton-neutron transition (kaon meson field) is coherent with the **SU (2)** symmetry of background transitions. The mechanism of commutation of left handed doublets and right handed singlet $2E_s=E_{act}$ (6) is similar to the **SU (2) xU (1)** symmetry group of fermions. At Eigen value, one the strong interaction of the quarks is confined.

The **2:1** flavor interactions (9) show that the kaon field is necessary to realize **SU (2)** symmetry and parity conservation within proton-neutron discrete transformations to realize discrete parity conservation within **uud-duu** and **ssc-ccs** nucleons.

Here is necessary to explain why three generation of particles, particularly strange (**s**) and charm (**c**) quarks are needed in particle physics[11]. Before we showed that if, correlation of different events with the energy resources has the same coupling they are simultaneous in time having different space phase. The invariance of Eigen value needs simultaneity in different phases. Therefore, identity conservation of the nuclear requires its discrete invariance within different space phases, formed by discrete existence of **uud-duu** and **ssc-ccs** frames.

Recently was observed that quarks treat the right and left hand differently and pick out a direction in space while in empty space they treat the left and right hand without any difference. On this basis, the strong interaction should violate the parity as well. Model (6) explains this phenomenon and shows that strong interactions can hold the parity only through discrete energy conservation while in the absence of constant coupling the strong interaction has to select one-handed direction.

The discrete energy conservation and action-response parity are the necessary laws of nature to give different shapes to the different events: without discrete conservation of energy and non-invariant action all the events and bodies would form non separable mass without any shape and structure.

The action principle (6) shows that the certain frame to be stable its outcome after the change should be the same (Eigen value **λ=1**). Therefore, formation of stable non-virtual matter frame needs application of additional intermediate force. Due to the conservation of energy in discrete mode, the intermediate force is necessary to recycle the energy to the nuclear.

It is known that the square of the parity transformation is the identity conservation. Applying two parity transformations is equivalent to no transformation. Therefore, the condition (6) meets the requirement of identity conservation generating discrete invariance of action-response parity. Model (6) involves non-invariance action instead of Yang-Mills gauge Lagrangian.

Model (6) shows that at constant Eigen value quark cannot radiate energy (**$E_s < E_{act}$**). Decay of proton by radiation of energy can be realized at inversion of the universe back to the initial state (**λ=-1**). In accordance with the model (6) proton decay has to be observed globally when **E_{act}=0.** This is the mathematical proof why quarks can never be elaborated from the hadrons.

By conventional particle theory when matter and antimatter collides, they should destroy each other, leaving behind nothing but only energy. In this concept energy as the resulting quantity is not localized with the lost of matter carrying momentum. That is why matter/antimatter annihilation should produce again other matter/antimatter couples to conserve momentum and space-time frame.

In accordance with discrete energy conservation, the anti-quark is the discrete partner of the quark and they exist with asymmetric boundaries. At small scale, the asymmetry is close to the antimatter structure, while for high space scale it is asymmetric structure of matter. The top quark's frame is close to the antimatter and it has no baryon structure. Due to the mechanism (7-16), during coupling of top-anti top quarks, individual bottom and anti bottom and vector bosons have not been observed [12]. The excess jet during proton-antiproton annihilation is due to the exchange of quarks. It is interesting that the possibility of generation of mass through top quark was suggested also by Liss and Tipton [12].By their opinion Higgs can be replaced by top-anti top quarks and there may be some heavy particle that decay to top-anti top pairs and excess jet is caused by collision of small particle with the top quarks.

By our model this heavy particle is not-yet described **[T-B]** bi-meson which in accordance with the mechanism (9-11) leads to the realization of **2:1** space-time performance of nucleons.

Model (6) predicts the non-symmetrical decay of top-anti top quarks. During correlation of top quark with up quark, it prefers to travel in direction of proton while anti top quark moves backward to the initial background state. One quark of baryonic 2:1 structure has different position in matrix (E_{act}/E_{s}-E_s/E_s) due to the correlation and insertion of energy to the baryonic structure.

At free state, photons themselves do not carry charge, while coupling with the charges makes them "color" force mediators (called gluons) which interact among themselves. Gluons carry the colors of their original photons. The color is the flavor of gluons while leptons carry the flavor of merged particles. The meson field of baryon having zero spin follows to the condition $E_{act=}E_s$ which keeps the color force constant as the quarks within baryon are pulled apart.

12. Does weak interaction violate the CP symmetry?

Landau [13] considered that parity violation is impossible because space is mirror symmetric and homogenous. In accordance with the discrete energy conservation, the spatial coordinate transformation is associated with the inversion of time and asymmetry of boundaries of space-time variables leads to the CP violation. The non-invariance of action –response generates discrete reflection.

W bosons as the intermediate products participate in the exchange of quarks within more massive top-bottom meson field. Therefore W bosons during coupling carry strong force while in the absence of $\mathbf{E}_{act}$ undergoes to the left-handed weak decay. When $\mathbf{E}_{act}$=0, takes place back re-alignment of $\mathbf{e}^+/\nu_e^+\mathbf{e}^-/\nu_e^-$to $\mathbf{e}^-/\mathbf{e}^+$+$\nu_e/\nu_e^-$ where the negative pressure dissolves $\mathbf{e}^-/\mathbf{e}^+$within ν_e/ν_e^-. That is why electron can behave in both left handed and right handed states.

One of the problems of Standard model is the CP invariance of strong interactions while quantum mechanics Lagrangian involves a term, which should break this symmetry.

Koboyashi and Maskawa's mechanism of CP violation [14], based on mathematical structure of quark matrix mixing, does not explain the physical nature of this violation.But quarks matrix mixing in reality is the change of space-time structure of the quarks in baryons.

In accordance with the discrete energy conservation concept the CP violation observed first time in Chui experiments [15] and repeated in beta decay, is due to the non-invariance of the action correlated with of the non-virtual space-time frame, therefore would result the CP violation.

As follows from the action-effect non-invariance parity (6), the left-right invariance can be realized only within discrete dynamics. The interaction described by **(V-A)/A** coupling is invariant only at λ=1.Therefore, parity cannot be conserved in the continuous symmetry.

13. Merging of quantum mechanics to the frame of classic physics

In accordance with the discrete energy conservation, model (6) as a quantum wave function may have a feature of deterministic classic physics and describe present existence of a particle in space-time frame through wave function connecting discrete squared appearance-disappearance phases. The probability of existence of a particle in one position is its discrete appearance in the space phase, while the non-existence describes its appearance within opposite time phase. Coupling of these phases in the wave function of probability is the "quantum analog" of the Eigen value. Therefore, with the discrete energy conservation concept there is no difference in the features of classic physics and quantum mechanics. Problem of quantum mechanics is that the wave function involves Hamiltonian of a single system having independent existence. On this basis, position and momentum are related to the system having an independent existence while in the case of discrete energy conservation concept position and momentum are derivatives of energetic existence. Description of existence of a single free particle makes the Hamiltonian linear operator eliminating its feature being the resulting quantity. Due to the independent existence of a system, Hamiltonian leads to the approximate solutions.

The non-independent existence differentiates small and high scale systems connecting them with the discrete energy conservation principle. At $\mathbf{E}_{act} \neq 0$ there is an existence of a particle in space-time frame displayed by its position and momentum while at $\mathbf{E}_{act=}0$ there is no position and momentum to be determined. Similar to our model (6), Hamiltonian as the Eigen value of the energy operator, in relation to the energy resources may present unification of relativity and quantum mechanics in function of discretely conserved field.

14. What special features has discrete energy conservation concept?

Our concept shows that the mathematical behavior of the non-continuous dynamical systems, described presently by ordinary differential equations, should be determined by the function involving commutation of the differential operator with the input function itself to cover the boundary of the phase of the function and identity conservation. Due to the discrete energy conservation mapped within discrete space-time boundary, all physical parameters are the resulting quantities.

The description of the " change" phenomenon by independent intervals of classic differentiation in the form of "continuous move from one point to another "gives only positive sign (such as Fermi's golden rule) and leads to the lost of discreteness and reversibility of dynamical events due to the elimination of the boundary of the initial function from the commutation. In this case, the identity conservation also is smeared out due to the absence of conjugation of the "energetic jump" with the dynamic boundary of space-time variables.

With the similar way, the second law of thermodynamics smeared out the discrete energy conservation by replacing discrete space-time dynamics by the one parametric thermodynamic arrow.

15. Conclusion

The analysis presented in our chapter shows that the concept of discrete energy/momentum conservation and their commutation within boundary mapped discrete space-time phases allow unification of the forces and interactions within unified classic field theory which completely changes our views on the fundamental interactions and symmetrical laws of nature.

The space and time are the products of discrete energy/momentum conservation and in reverse order, energy/momentum identities of antimatter/matter are the inner products of space-time discrete dynamics [16-18].

Unification of discrete energy conservation and discrete space-time dynamics has the same basis as electromagnetic unification of light as the coupling product of the space-time phases.

The principles that everything is "relative" or "uncertain " are getting different look with the concept that "every space-time identity is the discretely conserved energy/momentum packet within space and time phases" which became a new fundamental concept for description of nature and its physical laws regardless of dimensions and scale.

Model (6) shows that the non-conservation of parity is the result of non-arbitrary process of discrete energy conservation therefore the local CPT non-invariance is to be the fundamental deterministic law of nature. The discrete energy conservation treats its ordered outcomes-space and time through correlation of their asymmetric boundaries which allows energy to perform its discrete conservation within dynamic space-time framework. The non-invariance of action-response parity is more fundamental concept than unification of forces due to the generation of all the forces and events from discrete commutation of this relation.

The reality is not created by observation, as quantum mechanics suggests, but as the resulting quantity, it is created from exchange action-response interactions of space-time frame with the discrete energetic action.

Author details

Aghaddin Mamedov*

Address all correspondence to: amamedov@americas.sabic.com

SABIC Technology Center, Sugar Land, TX, United States

References

[1] P. Scarruffi. The nature of consciousness. 2006

[2] http://wikipedia. org/wiki/calculus, Access date missing (format:Accessed on dd/mm/yyyy)

[3] Dirac, P. A. M. (1958 (reprinted in 2011)). *Principles of Quantum Mechanics* (4th ed.). Clarendon. p. 255

[4] Dirac, P. A. M. (1928). "The Quantum Theory of the Electron". *Proceedings of the Royal Society A: Mathematical, Physical and Engineering Sciences* 117 (778):

[5] Vladimirov, V. S., Mikhailov, V. P., Vasharin A. A., et al., *Collection of Problems on Mathematical Physics Equations* [in Russian], Nauka, Moscow, 1974.

[6] Hazewinkel, M. Klein–Gordon equation", *Encyclopedia of Mathematics*. 2001

[7] P. Feynman. Nobel Lecture. 1965

[8] P. Pantell, H. E. Putoff. Fundamentals of Quantum Elrectrodynamics, New York, 1969

[9] S. Profumo. Lectures for TASI 2012, University of Colorado, June, 2012;ArXiv:1301. 0952v1, 5 Jan 2013

[10] H. Yukawa. Meson theory in its development. Nobel Lecture. 1949

[11] M. Gell Mann. Nobel lecture. Symmetry and currents in particle physics. 1969

[12] T. M. Liss, P. L. Tipton. The Discovery of Top Quark. Scientific American. Sep 1997

[13] L. D. Landay, and E. M. Lifshitz. Quantum Mechanics (Non-relativistic Theory) (Beijing World Publishing Corporation: P290 (1999)2.

[14] M. Koboyashi and T. Maskawa. Nobel lecture. 2005

[15] T. D. Lee. Nobel Lecture. 1957

[16] A. K. Mamedov. European Journal of Scientific Research. v. 36. No. 4. P. 570-584. 2009

[17] A. K. Mamedov. European Journal of Scientific Research. v. 42. No. 3. P. 359-384. 2010

[18] A. K. Mamedov. European Journal of Scientific Research. v. 57. No. 2. P. 314-365. 2011

7

The Computational Unified Field Theory (CUFT)–Revising Quantum and Relativistic Models

Jehonathan Bentwich

1. Introduction

1.1. The Computational Unified Field Theory (CUFT)

Over the past three years, a new hypothetical 'Computational Unified Field Theory' (CUFT) has been discovered which sets to unify between Quantum Mechanics and Relativity Theory (e.g., whose current theoretical contradiction is considered to be most likely the greatest unresolved enigma in modern Science). Indeed, several previous articles have demonstrated that this new hypothetical CUFT is capable of resolving the principle quantum-relativistic theoretical inconsistencies, replicating all of their key empirical phenomena, and was able to identify (three) "differential-critical" predictions differentiating it from both quantum and relativistic models of physical reality; Indeed, before proceeding to describe a (recent) empirical validation of one of these three CUFT 'differential-critical' predictions it may be helpful to delineate the key theoretical postulates underlying the CUFT as well as its associated "'Cinematic-Film Metaphor";

1.1.1. The 'Duality Principle'

The first theoretical postulate underlying the CUFT is the computational **'Duality Principle'** [4] which identified a basic "computational flaw" associated with both Quantum and Relativistic computational systems; The Duality Principle demonstrates that both quantum and relativistic computational systems comprise a 'Self-Referential Ontological Computational System' (SROCS) which assumes that it is possible to determine the value of any given 'y' (e.g., subatomic 'target' or relativistic 'space-time' or 'energy-mass' entity) strictly based on its direct (or indirect) physical interaction with another (exhaustive) 'x' factor/s (e.g., subatomic 'probe' or relativistic observer). It proves that such SROCS computational structure inevitably leads to both "logical inconsistency" and "computational indeterminacy" which are contradicted by robust empirical evidence indicating the capacity of both quantum and relativistic to determine the particular value of any given subatomic 'target' element or relativistic 'space-time' or 'energy-mass' phenomenon. Hence, the 'Duality Principle' negates the assumed SROCS computational structure underlying both quantum and relativistic computational systems, instead pointing at the existence of a (singular) higher-ordered 'Universal Computational Principle' (UCP) which computes the "simultaneous co-occurrence" of all exhaustive subatomic 'probe-target' or relativistic 'observer- (space-time or energy-mass) phenomenon' physical interactions (at any given point in time).

Indeed, the identification of such a singular **'Universal Computational Principle' (UCP)** responsible for the computation of all quantum and relativistic (exhaustive) physical interactions which is also postulated to possess three **'Computational Dimensions'** constitutes the second theoretical postulate of the CUFT; In order to perhaps better sense these three *'Computational Dimensions'* of the UCP let us examine a closely related **"Cinematic-Film Metaphor"** which may be used to explain these three 'Computational Dimensions' (as well as some other features and aspects of the CUFT):

Imagine yourself sitting in a cinema film presentation (e.g., seeing a film for the first time – unaware of the 'mechanics' of a film being presented to you)... In this case you could measure (for instance) the "velocity" (or energy) of a jet-plane zooming through the screen, the "time" it took this jet-plane to get from point 'A' to point 'B' (on the screen), the "spatial" length of the plane etc. – being unaware that (in truth) all of these 'spatial', 'temporal', 'energy' (and 'mass') "physical" features are produced based on the 'higher-ordered' computation of the degree of "displacement" or "lack of displacement" occurring across the series of cinematic-film frames!? Thus, for instance, the plane's "energy" (or velocity) is computed based on the number of 'pixels' that plane has been displaced across a given series of frames... Conversely, the plane's "spatial" measure is give based on the computation of the number of 'spatial pixels' that remain constant across a series of cinematic film frames (e.g., resulting in the fact that the plane's length doesn't "increase" or "decrease" across these frames)... Likewise, the "temporal" length of the plane's flight is computed based on the number of changes that occur in- or around- the plane (across a given number of film frames): imagine for instance what would happen to that plane's flight temporal value if the frames were projected more slowly (e.g., in "slow-motion" where there is a smaller number of changes taking place in the plane's flight, giving rise to a "dilated time" measure) or in a case in which precisely the same frame was presented over and over again for say one minute – time would "stand-still"... Similarly, we can devise a special 'cinematic-film' operation in which any given object is projected at "below-threshold" intensity at any given single frame such that only the presentation of the same object (in the same spatial configuration) across multiple number of frames may produce a visible object and that its apparent "mass" value will be computed as a function of the number of frames in which that object appeared 'spatially-consistent'... So, we can see that at least in the "cinematic-film metaphor", 'energy', 'space'; 'time' or 'mass' – are all produced as secondary computational measures being computed by a higher-ordered (singular) computation relating to the degree of 'changes'- or 'lack of changes'- of a given object across the frame, or as

Quite similarly, the CUFT posits that the four basic physical features of 'space', 'time', 'energy' and 'mass' are produced through the computation of a singular (higher-ordered) 'Universal Computational Principle' (represented by the Hebrew letter "yud") – of the degree of 'consistency' or 'inconsistency' across a series of extremely rapid (c^2/h) 'Universal Simultaneous Computational Frames' (USCF's): According to the CUFT, this Universal Computational Principle (UCP) employs two 'Computational Dimensions' to compute these four (secondary computational) physical features of 'space', 'time', 'energy' and 'mass' which are: 'Consistency' ('consistent' vs. 'inconsistent') and 'Framework' ('frame' vs. 'object'), and an additional Computational Dimension of 'Locus' ('global' vs. 'local') which accounts for relativistic phenomena.

1.1.2. The UCP's computational dimensions

Hence, the CUFT hypothesizes that the above mentioned 'Universal Computational Principle' (UCP) possesses three 'Computational Dimensions': The *'Framework'* Dimension relates to certain 'computational features' that are computed at the 'object' level, or at the 'frame' (USCF's) level; The 'Consistency' Dimension relates to the UCP's computation of the degree of 'consistency' or 'inconsistency' of an object across a series of USCF's frames (e.g., regarding its above mentioned 'object' or 'frame' measures and also relating to the below mentioned 'Locus' Dimension computation); and the 'Locus Dimension' relates to the UCP's computation of any 'Framework-Consistency' combination from computational perspective of the 'frame' (termed: 'global') or from the 'object's' computational perspective (termed: 'local'); The fascinating facet of these UCP's three Computational Dimensions is that they produce the four physical features of 'space', 'energy', 'mass' and 'time' – i.e., as secondary computational combinations of the 'Framework' and 'Consistency' Computational Dimensions: The CUFT posits that 'space' and 'energy' emerge as a result of the UCP's computation of the degree of 'consistent' or 'inconsistent' measure of an 'object' (e.g., comprising one of the computational levels of the 'Framework' Dimension) the 'Framework' Dimension; Likewise, the basic physical features of 'mass' and 'time' arise as secondary computational features associated with the degree of 'consistent' or 'inconsistent' measure of an object relative to the 'frame' (also comprising the 'Framework' Dimension)!

Hence, the (new) computational definitions of 'space', 'energy', 'mass' and 'energy' are given by:

S:(fi{x,y,z}[USCF(*i*)]+... f*j*{x,y,z}[USCF(*n*)]) / h x n{USCF's}

such that:

fj{x,y,z}[USCF(i)]) ≤ fi{x+(hxn),y+(hxn),z+(hxn)}[USCF(i...n)]

where the 'space' measure of a given object (or event) is computed based on a *frame consistent* computation that adds the specific USCF's (x,y,z) localization across a series of USCF's [1...n] – which nevertheless do *not exceed* the threshold of Planck's constant per each ('n') number of frames (e.g., thereby providing the CUFT's definition of "space" as 'frame-consistent' USCF's measure).

Conversely, the 'energy' of an object (e.g., whether it is the spatial dimensions of an object or event or whether it relates to the spatial location of an object) is computed based on the *frame's differences* of a given object's location/s or size/s across a series of USCF's, divided by the speed of light 'c' multiplied by the number of USCF's across which the object's energy value has been measured:

$$\mathbf{E:(f\mathit{j}\{x,y,z\}[USCF(n)])-f\mathit{i}\{(x+n),(y+n),(z+n)\}\ [USCF(\mathit{i...n})]\ )\ /c \times n\{USCF's\}}$$

such that:

$$fj\{x,y,z\}[USCF(n)])>fi\{x+(hxn),y+(hxn),z+(hxn)\ [USCF(i\ldots n)])$$

wherein the energetic value of a given object, event etc. is computed based on the subtraction of that object's "universal pixels" location/s across a series of USCF's, divided by the speed of light multiplied by the number of USCF's.

In contrast, the of 'mass' of an object is computed based on a measure of the number of times an *'object'* is presented *'consistently'* across a series of USCF's, divided by Planck's constant (e.g., representing the minimal degree of inter-frame's changes):

$$\mathbf{M:\ \Sigma[o\mathit{j}\{x,y,z\}\ [USCF(\mathit{n})] = o(\mathit{i\ldots j\text{-}1})\ \{(x),(y),(z)\}\ \{\mathit{USCF(i...n)}\}\ /\ h \times n\{USCF's\}\ \{\mathit{USCF(1...n)}\}\ /\ h \times n\{USCF's\}}$$

where the measure of 'mass' is computed based on a comparison of the number of instances in which an object's (or event's) 'universal-pixels' measures (e.g., along the three axes 'x', y' and 'z') is identical across a series of USCF's (e.g., $\Sigma oi\{x,y,z\}\ [USCF(n)] = oj\{(x+m),(y+m),(z+m)\}\ [USCF(1...n)]$), divided by Planck's constant.

Again, the measure of 'mass' represents an *object-consistent* computational measure – e.g., regardless of any changes in that object's spatial (frame) position across these frames.

Finally, the *'time'* measure is computed based on an *'object-inconsistent'* computation of the number of instances in which an 'object' (i.e., corresponding to only a particular segment of the entire USCF) changes across two subsequent USCF's (e.g., $\Sigma\ o\mathit{i}\{x,y,z\}\ [USCF(\mathit{n})] \neq o\mathit{j}\{(x+m),(y+m),(z+m)\}[USCF(\mathit{1...n})]$), ivided by 'c':

$$\mathbf{T:\Sigma\ o\mathit{j}\{x,y,z\}\ [USCF(\mathit{n})] \neq o(\mathit{i\ldots j\text{-}1})\{(x),(y),(z)\}\ [USCF(\mathit{1...n})]\ /c \times n\{USCF's\}}$$

such that:

$$T{:}\Sigma o\mathit{i}\{x,y,z\}[USCF(\mathit{n})]\text{-}\mathit{j}\{(x+m),(y+m),(z+m)\}\ [USCF(\mathit{1...n})] \leq c \times n\{USCF's\}$$

Hence, the measure of 'time' represents a computational measure of the number of *'object-inconsistent'* presentations any given object (or event) possesses across subsequent USCF' (e.g., once again- regardless of any changes in that object's 'frame's' spatial position across this series of USCF's). Finally, the combination of the 'Locus' Dimension together with the 'Framework-Consistency' Dimensions, e.g., producing the four physical features of 'space', 'energy', 'mass', and 'time' – produces all known relativistic effects and phenomenon, e.g., such as 'time-dilation', 'energy-mass' equivalence and even the curvature of 'space-time'!

1.1.3. The Computational Invariance Principle

Another key theoretical postulate comprising the CUFT is the 'Computational Invariance Principle' which identifies this 'Universal Computational Principle' as the sole 'computationally invariant' element which both produces all four 'computationally variant' physical features of 'space', 'time', 'energy' and 'mass' and also exists independently of these physical features "in-between" any two subsequent 'USCF's frames; As such, the 'Computational Invariance Principle' recognizes the Universal Computational Principle as the sole (and singular) 'invariant' reality underlying the production of the four secondary computational 'variant' physical properties of 'space', 'time', 'energy' and 'mass' (based in part on a well-known scientific principle: "Ockham's Razor" which prefers the simplest most parsimonious theoretical account for complex phenomena) [1]...

1.1.4. The Universal Computational Formula

Finally, this recognition of the Universal Computational Principle as the sole and singular reality producing and sustaining all four (secondary computational) physical properties of space, time, energy and mass has also lead to the formulation of a singular 'Universal Computational Formula' which completely integrates these four secondary computational physical properties, as well as all known quantum and relativistic properties (e.g., as embedded within the higher-ordered Universal Computational Formula):

$$\text{Universal Computational Formula}: \left\{ \frac{`c^2}{h} \right\} = \frac{s}{t} \times \frac{e}{m}$$

2. The CUFT: Quantum-relativistic harmonization- embedding- & transcendence

Hence, the next necessary step in validating the CUFT as a satisfactory TOE is to demonstrate that it's capable of harmonizing between quantum and relativistic models, embed both models within the CUFT's Universal Computational Formula, and providing certain "differential critical predictions" which transcend these quantum and relativistic models (e.g., and if validated empirically validate the CUFT as an satisfactory TOE!) First, we set to demonstrate that the CUFT is able to bridge the (above mentioned) key theoretical inconsistencies that seem to exist between quantum and relativistic models based on its reformulation- and embedding- of quantum and relativistic computation within the singular (higher-ordered) Universal Computational Principle (e.g., due to the Duality Principle's identification of a mutual computational flaw underlying both models, as shown above); Interestingly, based on the singularity of the Universal Computational Principle' computation of both quantum and relativistic relationships (e.g., as embedded within an exhaustive USCF's frames' series) the CUFT is able to embrace both quantum's probabilistic and positivistic relativistic modeling. This is because the Universal Computational Principle's (rapid) production of all exhaustive spatial-pixels in the physical universe comprising each USCF frame – allows it to embed "single spatial-temporal" relativistic objects' (or subatomic 'particles') measurements as well as "multi spatial-temporal" subatomic 'wave' measures! In fact, one of the elegant features of the CUFT is precisely the fact that it conceptualizes such 'single spatial-temporal' relativistic 'objects' or subatomic quantum 'particles' and 'multi spatial-temporal' quantum subatomic 'wave' measurements – within the exhaustive computational framework of the Universal Computational Principle's rapid production of the series of USCF's frames (e.g., comprising all such 'single spatial-temporal' relativistic object or subatomic particle and 'multi spatial-temporal' quantum wave measurements...) Moreover, this exhaustive computational framework of the CUFT allows it to reconceptualize quantum's Uncertainty Principle's 'complimentary pairs' of 'space and energy' or 'time and mass' merely representing a computational constraint intrinsically embedded within the Universal Computational Principle's computation of the two 'Framework' and 'Consistency' Computational Dimensions – i.e., based on the fact that 'space' and 'energy' exhaustively comprise the Framework's Dimensions' 'frame' level, and likewise 'mass' and 'time' exhaustively comprising Framework's 'object' computational level... Hence, the CUFT is capable of embedding both 'single spatial-temporal' relativistic objects (and quantum 'particles'), and (apparently) 'probabilistic' 'multi spatial-temporal' quantum wave measures within the broader and more exhaustive Universal Computational Principle's rapid computation of the series of USCF's frames (e.g., thereby also resolving the 'particle-wave duality' postulate of Quantum Mechanics!)

The CUFT's resolution of the second key quantum-relativistic theoretical inconsistency relating to quantum's instantaneous 'entanglement' phenomenon as opposed to Relativity's speed of light constraint set on the transmission of any signal across space is also anchored in the above mentioned Universal Computational Principle's rapid computation of these USCF's frames ; Since the CUFT posits that the Universal Computational Principle's (rapid) computation of each of the Universal Simultaneous Computational Frame (USCF) simultaneously computes all of the spatial-pixels in the physical universe at a minimal time-point (e.g., c^2/h), then this computation extends the phenomenon of 'quantum entanglement' to all exhaustive spatial points in the universe (e.g., at any such minimal time-point! On the other hand, based on the above embedding of all 'single spatial-temporal' relativistic objects (or subatomic particles) as well as 'multi spatial-temporal' subatomic wave measures within the Universal Computational Principle's exhaustive USCF's computation – it allows for Relativity's apparent speed of light constraint imposed on any such 'single spatial-temporal' relativistic object (or subatomic 'particle') transmission!

The next step towards the validation of the CUFT as a satisfactory TOE involves an articulation of the embedding of quantum and relativistic models within the singular higher-ordered CUFT's Universal Computational Formula – i.e., which is shown to both maintain- and transcend- the (currently) known quantum and relativistic relationships! As can be seen from the two 'quantum' and 'relativistic' formats of the Universal Computational Principle (below), the highlighted portions of these formats conforms to the known mathematical relationships found in quantum and relativistic models, e.g., Relativity's energy and mass equivalence, and Quantum's 'complimentary pairs' of 'space and energy', 'mass and time' as constrained by the Uncertainty Principle's 'h' Planck's constant simultaneous measurement accuracy constraint:

I. Relativistic Format: $e \times \frac{s}{t} = m \times \frac{c^2}{h}$

II. Quantum Format: $t \times m \times \frac{c^2}{h} = s \times e$

3. The CUFT's "differential-critical predictions"

However, it also becomes clear that the CUFT's Universal Computational Principle's embedding of those empirically validated quantum and relativistic relationships – also transcends and critically differs from these relationships! Indeed, these computational differences between the Universal Computational Formula's 'quantum' and 'relativistic' formats and the 'standard' relativistic '$E=Mc^2$' and quantum 'complimentary pairs' constitutes one (of three) "differentia-critical predictions" that differentiate the CUFT model from both quantum and relativistic predictions, e.g., thereby providing an empirically testable means for validating the CUFT as a satisfactory TOE...

Another key "differential-critical prediction" that differentiates the CUFT from both relativistic and quantum models' predictions are: the CUFT's prediction regarding the more consistent (spatial) presentation of more massive particles (or elements) – across a given series of USCF's frames, relative to less massive particles' appearance across the same series of USCF's frames. In fact, this 'differential critical prediction' regarding the more consistent spatial presentation of more massive particles (or elements) across a series of USCF's frames, relative to the spatial presentation of less massive particles (or elements) precisely replicates the empirical findings of the recently discovered 'Proton-Radius Puzzle', thereby providing a first empirical validation for the CUFT as a satisfactory TOE!

The third 'differential-critical prediction' differentiating the CUFT from both quantum and relativistic models involves a possible "reversal of the space-time spatial-electromagnetic pixels sequence" across a series of USCF's frames electromagnetic spatial-pixels' sequence of a given object or phenomenon; This may be achieved through a precise recording of that object (or phenomenon's) spatial-electromagnetic pixels values (across a given series of USCF's), and a manipulation of these electromagnetic-spatial pixels values (through precise electromagnetic stimulation) so as to produce the reverse sequence of the recorded spatial-electromagnetic values sequence! Interestingly, due to the fact that quantum theory precludes the possibility of the "un-collapse" of the probability wave function following a certain interaction between the any such probe particle and the target particle's wave function – this 'differential critical prediction' is ruled out as a possible prediction of Quantum Mechanics; Likewise, since Relativity sets the speed of light as a clear "unsurpassable" limit for the transference of any signals it also precludes the possibility of "reversing time"; In contrast, since the CUFT defines 'time' (e.g., alongside the other three physical features of 'space', 'energy' and 'mass') merely as a secondary computational property produced by the Universal Computational Principle's three Computational Dimensions' computation of the degree of an "object's-inconsistency" across a series of USCF's frames – then it should allow for the "reversal" of the 'space-time' sequence (e.g., of the particular spatial-electromagnetic pixels' values) across a series of USCF's frames!

4. Empirical validation of the CUFT as satisfactory 'TOE': The 'proton radius puzzle'

Fortunately, the second (abovementioned) 'differential-critical prediction' of the CUFT regarding the more consistent spatial presentations of a more massive particle (or element), relative to the spatial-consistency of a less massive particle (or element) across a given series of USCF's frames – has now received initial empirical validation through the findings associated with the 'Proton-Radius Puzzle'! This is because the 'Proton-Radius Puzzle' empirical findings indicate that the more massive 'Moun Hydrogen Proton' is measured (approximately) 200 times – smaller and more accurate than the standard Hydrogen (e.g., with the 200 times lighter electron particle instead of the Muon)... In order to fully understand how these 'Proton-Radius Puzzle' findings (Bernauer & Pohl, 2014) empirically confirm the differential-critical prediction of the CUFT, lets us return to the CUFT's computational definitions of "mass"; Mass is defined by the CUFT as a measure of the degree of "spatial-consistency" of a particle across a given series of USCF's frames! In mathematical terms, it is measured as the number of times that this particle was presented across the same spatial pixels (measured from within the object's frame of reference) across a series of USCF's frames... This computational definition of 'mass' implies at least two empirically measurable predictions:

a. That the more massive 'Muon' particle should be measured as more accurate- and as smaller- than the less massive electron particle; this is due to the fact that the more massive a particle is the greater its spatial-consistency across USCF's frames! And/or:

b. That more massive particles (e.g., such as the Muon) should be measured across a greater number of USCF's frames, relative to less massive particles (such as the electron); In other words, we could expect to measure the (more massive) Muon across a greater number of USCF's frames than the (lighter) electron!

Interestingly, the 'Proton-Radius Puzzle' precisely confirms the first of these two CUFT 'differential critical' predictions – i.e., indicating that the (200 times) more massive Muon particle (e.g., when embedded within the Hydrogen Proton) is measured as (200 times) 'smaller' and 'more accurate' than the (200 times) less massive electron (associated) Hydrogen Proton! Hence, these findings provide an initial empirical confirmation of the CUFT – as differing from the predictions of both quantum and relativistic models' predictions (e.g., which cannot account for these "Proton-Radius Puzzle" findings)!

Efforts should be made to empirically validate the second (abovementioned) aspect of the CUFT's differential-critical prediction regarding the appearance of 'more massive' particles such as the Muon across a greater number of USCF's frames than the appearance of less massive particles (such as the electron)!

5. The CUFT: Challenging quantum & relativistic "materialistic-reductionistic" assumption

Thus far, we've been able to demonstrate that the CUFT may be considered a satisfactory 'TOE' capable of resolving all major quantum-relativistic theoretical inconsistencies, replicating their primary empirical phenomena, identifying and empirically validating one of the CUFT's 'differential-critical' predictions differentiating it from both quantum and relativistic predictions... The primary aim of the current manuscript is to utilize this recognition of the CUFT as a satisfactory TOE, e.g., which also embeds both quantum and relativistic models within its broader more comprehensive (singular) 'Universal Computational Principle' theoretical framework – *towards recognizing the need to revise certain key theoretical aspects of both quantum and relativistic fields, i.e., based on the CUFT's singularity of the UCP sole production of the (extremely rapid: c^2/h) series of the 'Universal Simultaneous Computational Frames' (USCF's); Specifically, the CUFT's emphasis on the singularity of the UCP (rapid) production of the USCF's series – forces us to revise both Quantum and Relativistic "materialistic-reductionistic" basic assumption whereby any (quantum or relativistic) physical relationship (or entities, value/s, phenomenon) can be determined solely based on an exhaustive probe-target (subatomic) interaction or observer-phenomenon (e.g., space-time or energy-mass) relationship in such a manner as to point at the sole and singular production of all such quantum and relativistic entities, phenomena, relationship/s by the UCP's USCF's production...*

Indeed, if we revert back to the CUFT's (first) 'Duality Principle' theoretical postulate, we can see that both Quantum and Relativistic computational systems comprise a *'Self-Referential Ontological Computational Systems' (SROCS)* structure; this quantum and relativistic SROCS computational structure is synonymous with a **"materialistic-reductionistic" assumption**, wherein it is assumed that the determination of the "existence"/"non-existence" of any given 'y' entity (or value) is determined solely based on its direct (or indirect) physical interaction with another (exhaustive) 'x' factor/s... As we've seen, the Duality Principle in fact negates the validity of such assumed (quantum or relativistic) SROCS systems – instead, pointing at the existence of the singular higher-ordered (D2) 'Universal Computational Principle' (UCP) which alone computes the "simultaneous co-occurrence" of all (exhaustive) quantum and relativistic 'probe-target' and 'observer-phenomenon' pairs series (e.g., subsequently shown by the CUFT to comprise any minimal time-point 'Universal Simultaneous Computational Frame'). The CUFT further developed this 'Duality Principle' and 'UCP' (alongside its three 'Computational Dimensions') postulates towards the recognition of the *'Computational Invariance Principle'*: i.e., recognizing the fact that since only the UCP "exists" both 'during' each of the USCF's frames (in fact producing all of its exhaustive universal spatial pixels simultaneously at any such minimal time-point) as well as 'solely existing' "in-between" any two subsequent USCF frames (whereas the four secondary computational 'physical' features of 'space', 'time', 'energy' and 'mass' only exist "during" the UCP's production of the USCF's and its computation of these four secondary computational physical features), then we must conclude that only this singular UCP comprises an invariant "reality" (whereas these four secondary-computational 'physical' features may only be considered *'phenomenally' variant*)... Hence, the CUFT's 'Computational Invariance Principle' in fact points at the sole reality of the UCP (e.g., computationally invariant), as opposed to the "phenomenal" nature of the four secondary computational 'physical' features of 'space', 'time', 'energy' and 'mass' (e.g., computationally variant).

It is hereby suggested that a deeper analysis of these three particular theoretical postulates of the CUFT (e.g., the 'Duality Principle', 'Universal Computational Principle' and 'Computational Invariance Principle') may negate the current (quantum and relativistic) *"materialistic-reductionistic assumption"* (e.g., represented by the SROCS computational structure) based on the sole and singular reality of the 'Universal Computational Principle'; This is made particularly clear based on (above mentioned) 'Computational Invariance' Principle's proof for the singular reality of the Universal Computational Principle – which is the only 'computationally invariant' element which "exists" both during its sole production the (rapid series of) USCF's frames, and "in-between" any two such (subsequent) USCF's frames! This is because based on this 'Computational Invariance Principle', the four physical features of 'space', 'time', 'energy' and 'mass' constitute computationally 'variant' properties and are therefore 'transient' (i.e., exist only "during" the Universal Computational Principle's production of the USCF's frames but ceases to exist "in-between" any two such USCF's frames)... Indeed, their "computational variant" composition makes them possess only "phenomenal" validity as opposed to the singular reality of the 'Universal Computational Principle' which exists permanently (and solely) – both as producing these 'phenomenal' (computationally variant) four 'physical' features "during" the USCF's frames and also "in-between" these USCF's frames... Therefore, the sole production- sustenance- and "transference" of any of these four "physical" features – during or across – any USCF frame/s is only made possible through the singular existence (and operation) of the 'Universal Computational Principle'! In other words, since these four 'phenomenal-physical' features "exist" only during each USCF frame, e.g., as produced by the singular 'Universal Computational Principle', but not "in-between" any two such subsequent USCF's frames – as opposed to the singularity of the UCP which solely produces these four physical features "during" the USCF's frames and also exists "in-between" the USCF's frames; then, we must conclude that the only means for the "production"- "sustenance"- or "transference" of any given 'physical' feature across any two (subsequent) USCF's frames may only be done based on the UCP!

Hence, we must conclude that the basic assumption of "materialistic-reductionism", e.g., whereby it is possible to determine the "existence" or "non-existence" of any given 'physical' 'y' feature solely based on its direct or indirect physical interaction with another (exhaustive) 'x' factor/s – is negated (not only by the above mentioned Duality Principle) but even more explicitly through the recognition that it is not possible for any of these four (phenomenal) 'physical' features to be "transferred" across any two subsequent USCF's frames – except through the computation of the Universal Computational Principle, which constitutes the sole (computationally invariant) "reality" (which exists both "during" the USCF frames producing these four phenomena 'physical' features and solely exists "in-between" any two USCF's frames...) What this means is that the basic "materialistic-reductionistic" assumption underlying both quantum and relativistic SROCS computational systems, i.e., which assumes that the "existence" (or "non-existence") of any given subatomic 'target' or relativistic (space-time or energy-mass) -'phenomenon' is determined solely based on their direct or indirect physical interaction/s with another (exhaustive) subatomic 'probe' element or relativistic 'observer' – is negated! Instead, the 'production'- 'sustenance'- or 'development'- of any 'physical' feature (relationship or phenomenon) – at the quantum or relativistic frameworks can only be computed through the singularity of the Universal Computational Principle!

6. Revising physics: UCP a-causal computation

Hence, there seems to arise a necessity to revise both quantum and relativistic computational systems such that the 'existence' of any of the four (computationally variant phenomenal) physical features (of 'space', 'time', 'energy' or 'mass') in either quantum or relativistic theoretical frameworks be solely produced- sustained- or developed- solely based on the Universal Computational Principle's singular production of all spatial pixels in the universe at any minimal USCF frame/s time-point; Note, however, that this revision does not represent merely a 'philosophical' concept – i.e., in fact, it is suggested that this revision signifies a fundamental shift in Physics as it relies on the **UCP** singular (higher-ordered) **"A-Causal Computation"!**

In order to fully grasp the potential significance of this (novel) UCP 'A-Causal Computation', it may be helpful to specify the theoretical ramifications of recognizing the fact that in both Quantum Mechanics and Relativity Theory the sole production- sustenance- and development- of any of the four 'physical' (phenomenal) features can only be computed by the UCP; Given the fact that the UCP is postulated to compute *"simultaneous co-occurrence"* of all (exhaustive) quantum 'probe-target' and relativistic 'observer-phenomenon' relationships, this means that both within a *single USCF* frame and across a *series* of such *USCF's* frames – we cannot (any longer) rely on any 'materialistic-reductionistic' (SROCS) subatomic 'probe-target' or relativistic 'observer-phenomenon' interactions for determining any of the four (quantum or relativistic) 'physical' features... Instead, we must revise any such quantum or relativistic 'physical' feature based on the UCP computation of the "simultaneous co-occurrence" of all (exhaustive) quantum and relativistic relationships comprising any (single or multiple) USCF frame/s! Indeed, this fundamental shift from the current "materialistic-reductionistic" quantum or relativistic (SROCS) assumption towards a recognition of the sole computation of the UCP of all 'simultaneously co-occurring' quantum and relativistic ('probe-target' and 'observer-phenomenon') interactions is termed: the 'UCP A-Causal Computation' (i.e., of all 'simultaneously co-occurring quantum and relativistic relationships comprising any single or multiple USCF's)...

Now, in order to understand the far reaching theoretical implications of recognizing this UCP (singular) higher-ordered 'A-Causal Computation' let us turn our attention to the two (overarching) conceptual models of the "probabilistic interpretation of Quantum Mechanics" (i.e., represented by the 'probability wave function' and its "collapse" following any given subatomic probe measurement) and Einstein's (famous) General Relativity Einstein field equations (EFE): $R_{\mu\nu} - \frac{1}{2} g_{\mu\nu} R + g_{\mu\nu} \Lambda = \frac{8\pi G}{c^4} T_{\mu\nu}$ (describing the dynamic interaction that exists between massive object's curvature of the fabric of 'space-time' which in return determines their travelling pathway, and vice versa...) It is suggested that in both of these cases, their current theoretical formulation represents the above mentioned "materialistic-reductionistic" assumption – i.e., whereby it is the direct or indirect physical interaction/s between a given subatomic 'probe' and 'probability wave function' target element which determines the "collapse" of that wave function and hence the value of the measured 'target particle; or it is the direct physical interaction of a given (massive) object with the 'space-time' which determines its curvature – and this curvature of 'space-time' (in return) interacts with this given (massive) object thereby determining its pathway movement...

Indeed, the 'materialistic-reductionistic' structure of any such (hypothetical) quantum 'probe-target' or relativistic 'observer-phenomenon' relationships was analyzed earlier, and proven to comprise a SROCS computational structure, e.g., being negated by the CUFT 'Duality Principle' – pointing at the necessity to reformulate both quantum and relativistic computational systems based on the singularity of the Universal Computational Principle...

But, what becomes apparent here, e.g., based on the recognition of the UCP "A-Causal Computation" is that the basic (overarching) theoretical model of both Quantum Mechanics and (General) Relativity Theory must be revised based on this 'UCP A-Causal Computation'! This is because once we accept the CUFT assertion that there exists only one singular (computationally invariant) UCP "reality" and that this UCP singular 'reality' solely computes the "simultaneous co-occurrence" of all exhaustive (quantum and relativistic) relationships comprising a (minimal time-point) USCF frame/s (e.g., termed: 'UCP A-Causal Computation'), then we cannot (any longer) retain either QM's current model regarding the 'materisalistic-reductionistic collapse of the probability wave function' or General Relativity's EFE. This is because both QM's assumed (SROCS) collapse of the probability wave function, as well as RT's assumed SROCS (massive) object –space-time curvature computational structure – are based on this "materialistic-reductionistic SROCS' assumption; In the case of QM this 'materialistic-reductionistic' SROCS computational structure is represented in the assumption wherein the determination of the values of any subatomic 'target' (probability wave function) element is contingent upon its direct (or indirect) physical interaction with another subatomic 'probe' element – i.e., which "causes" the "collapse" of the probability wave function (see earlier description of the Quantum 'probe-target' SROCS structure and its violation of the Duality Principle); In the case of (General) Relativity Theory, this 'materialistic-reductionistic SROCS assumption' is represented by through Relativity's EFE which determines the 'curvature of space-time' based on its direct physical interaction with 'massive objects' and vice versa determines the movement of these 'massive objects' based on their interaction with the curvature of 'space-time' (also see the earlier Duality Principle's analysis of relativistic 'observer-phenomenon' SROCS computational structure);

However, based on the above recognition of the singular reality of the UCP which solely computes the "simultaneous co-occurrence" of all (exhaustive) quantum and relativistic interactions, i.e., comprising any single or multiple USCF's frames (e.g., termed: the UCP 'A-Causal Computation), we must revise this basic 'materialistic-reductionistic SROCS assumption' based on this higher-ordered singular UCP A Causal Computation! In other words, since the sole production- sustenance- and development- of any of the four (phenomenal) 'physical' features (e.g., say in the quantum domain) is based on the UCP 'A-Causal Computation', i.e., of all exhaustive 'probe-probability wave function target' interactions comprising a single (or multiple) USCF frame/s, then we must also conclude that the apparent "collapse" of the probability wave function – cannot be "caused" by the direct interaction between any given probe element and given probability wave function! This is simply due to the fact that according to the above 'UCP A-Causal Computation' all exhaustive values of all quantum subatomic 'probe-probability wave function target' interactions are *computed simultaneously by the UCP – comprising all spatial pixels comprising any single or multiple USCF frame/s...* And since the only "transference" of any of the four (phenomenal) 'physical' features from one USCF frame to another – can only be carried out through the singular operation of the UCP's production- sustenance- and development- of any spatial pixel in te universe (across USCF's frames), then we cannot attribute the "collapse of the wave function" to any physical interaction taking place between any 'probe' and 'probability wave function target' entities (e.g., at any particular USCF frame/s)... Hence, the UCP's A-Causal Computation which produces simultaneously all exhaustive quantum 'probe-target' interactions – at any single or multiple USCF frame/s negates the validity of the "materialistic-reductionistic SROCS" assumption of the "collapse of the probability wave function" (target element) as a result of its direct interaction with another subatomic 'probe' element.

Likewise, based on the recognition of the UCP's singular 'A-Causal Computation' which is solely responsible for the production- sustenance- and development- of all exhaustive 'observer – (space-time, energy-mass) phenomenon' interactions comprising any (single or multiple) USCF frame/s, we must revise the current EFE representing a "materialistic-reductionistic SROCS assumption"; Once again, this is due to the fact that contrary to this 'materialistic-reductionistic SROCS' assumption represented by Relativity Theory EFE according to the UCP's singular 'A-Causal Computation' the UCP computes the "simultaneous co-occurrence" of all exhaustive relativistic 'observer – (space-time or energy-mass) phenomenon' relationships comprising any (single or multiple) USCF frame/s! Therefore, contrary to (General) Relativity Theory's currently assumed 'materialistic-reductionistic SROCS' assumption, wherein the 'curvature of space-time' is determined through its direct physical interaction with 'massive objects' (and vice versa, the movement of these 'massive objects' is determined strictly based on the 'curvature of space-time') – the UCP's A-Causal Computation asserts that it is solely the singularity of the UCP which computes the "simultaneous co-occurrences" of all exhaustive relativistic 'observer – (space-time or energy-mass) phenomenon' interactions (e.g., comprising any single or multiple USCF/s)...

Hence, the fundamental necessary revision of both (contemporary) probabilistic interpretation of QM and of (General) Relativity Theory involves a shift from the current 'materialistic-reductionistic SROCS' assumption underlying both QM and RT – towards the UCP's singular 'A-Causal Computation'! Essentially, this revision implies that instead of the currently assumed (quantum or relativistic) 'materialistic-reductionistic SROCS' assumption wherein the "collapse of the target probability wave function' is "caused" by its direct physical interaction with the subatomic 'probe' element, and the 'curvature of space-time' is "caused" by its direct physical interaction with 'massive object/s' (and vice versa the movement of these 'massive object/s' is "caused" by the 'curvature of 'space-time'; the UCP 'A-Causal Computation' negates any such 'materialistic-reductionistic' "causal" relationships, instead pointing at the fact that it is only the singularity of the UCP which computes the "simultaneous co-occurrence" of all (exhaustive) quantum 'probe-target' and relativistic 'observer-phenomenon' interactions comprising any (single or multiple) USCF frame/s... Perhaps another (lucid) manner of demonstrating the UCP's negation of contemporary Quantum and Relativistic 'materialistic-reductionistic SROCS' computational structure can be given through an analysis of the minimal-temporal, i.e., USCF's frames dynamics representing Quantum Mechanics' currently assumed "collapse of the target probability wave function", as well of the USCF's frames' dynamics representing Relativity's EFE (e.g., describing the interactive effect of 'massive objects' on the 'curvature of space-time' and vice versa as explained above); According to the contemporary 'probabilistic interpretation of QM' the target's probability wave function "collapses" as a result of its direct physical interaction with another subatomic probe element: this means that at a particular minimal-time USCF frame there occurs a direct physical interaction between the 'target's probability wave function' and the 'probe element' – and that based on this direct 'probe-target probability wave function' physical interaction this 'target probability wave function' "collapses"; But since each (single) USCF frame comprises the "minimal time-point" (possible) at which the (singular) UCP produces all exhaustive (quantum and relativistic) spatial-pixels in the universe, then necessarily the initial direct physical interaction between the 'probe' element and the 'target's probability wave function' – takes place at a given USCF frame, whereas the (assumed) "resulting collapse" of this 'target's probability wave function' must occur at a subsequent USCF frame! But, since according to the above mentioned 'Computational Invariance Principle' the sole and singular (computationally invariant) principle which exists both "during" and "in-between" any two subsequent USCF frames is the UCP (whereas the four phenomenal 'physical' features of 'space', 'time', 'energy' and 'mass' exist only "during" any given USCF frame), then the sole 'production'-sustenance- and 'development' of any quantum (or relativistic) (phenomenal) 'physical' feature may only be carried out by the singularity of the UCP – which indeed computes the "simultaneous co-occurrence" of all exhaustive quantum (or relativistic) 'probe-target' interactions comprising any single or multiple USCF's... Hence, since it is not possible for any physical interaction (say) between the 'subatomic probe' and (assumed) 'target's probability wave function' at a given USCF frame (i) to have any effect on their (phenomenal) 'physical' features at a subsequent USCF frame (i+1) – but rather it is the sole (and singular) computation of the UCP of the "simultaneous co-occurrences" of any exhaustive 'probe-target' subatomic interaction/s (at any single or multiple USCF/s frames) which produces- sustains- and develops- any exhaustive subatomic probe-target relationships! Likewise, we can show that based on such 'minimal time-point' USCF's frames analysis, that the currently assumed 'materialistic-reductionistic SROCS' General Relativity Theory model's interactive 'curvature of space-time' based on its direct physical interaction with 'massive objects' (and vice versa) is negated – instead, pointing at the sole production- sustenance- and development- of any of the four phenomenal 'physical' features (of 'space', 'time', 'energy' and 'mass') including the phenomena of the curvature of 'space-time' or of the (apparent) movement of massive objects solely based on the singular UCP 'A-Causal Computation'; In order to demonstrate the impossibility of the currently assumed Relativistic 'materialistic-reductionistic SROCS' assumption, let us (once again) imagine the 'minimal time-point dynamics' of the currently assumed (General) Relativity 'materialistic-reductionistic SROCS' direct physical interaction between certain massive object/s and the curvature of space time (and vice versa, as outlined above): According to this 'materialistic-reductionistic SROCS' Relativistic assumption, it is the direct physical interaction that exists between (one or more) 'massive object/s' and 'space-time' which "causes" this 'space-time' (fabric) to curve, and vice versa, this 'curvature of space-time' "causes" any given 'massive object/s' to travel in a particular (curved) space-time pathway... But, when analyzed from the perspective of a 'minimal time-point' USCF frame/s, then we see that in the first (hypothetical) USCF frame, there is a direct physical interaction between the given 'massive object' and the (fabric of) 'space-time', whereas this 'curving of space-time' may affect the space-time movement pathway of this 'massive object' – only in a subsequent USCF frame/s... However, such 'materialistic-reductionistic' (SROCS) relativistic assumption – not only violates the Duality Principle (as shown earlier and previously), but in the context of this 'minimal time-point' USCF's frames analysis seems to negate the (proven) singularity of the UCP's 'A-Causal Computation': This is because, according to the above 'materialistic-reductionisitc' relativistic assumption it is the direct physical interaction between the given 'massive object' and the (fabric of) 'space-time' – in the first USCF frame/s which "causes" this (fabric of 'space-time') to curve in subsequent USCF frame/s, and this 'curvature of space-time' in turn "causes" this given "massive object" to travel in a "curved space-time pathway" in still later USCF frame/s... However, based on the UCP's 'A-Causal Computation', which computes the "simultaneous co-occurrences" of all exhaustive relativistic 'observer-phenomenon' relationships, we have to negate this relativistic 'materialistic-reductionistic SROCS' assumption – since there cannot exist any "cause and effect" relationship between any direct (or indirect) physical interactions (e.g., such as 'massive object/s' which 'curve the fabric of space-time') at an initial USCF frame/s, and its effect on the 'space-time movement pathway of that massive object' at a subsequent USCF/s... Instead, the singularity of UCP's 'A-Causal Computation' forces us to recognize its computation of the 'simultaneous co-occurrences' of all (four) phenomenal 'physical' features – i.e., including the simultaneous computation of any 'massive object' and any 'curvature of space-time'!

Hence, we reach the inevitable conclusion whereby both Quantum and Relativistic models have to be revised in terms of their basic 'materialistic-reductionistic SROCS' assumption, i.e., recognizing the fact that either Quantum Mechanics' assumed 'collapse of target's probability wave function' as 'caused' by its direct physical interaction with another subatomic 'probe' element; or Relativity's assumed 'curvature of space-time' as "caused" by its direct physical interaction with 'massive objects' – is negated by the CUFT's recognition of the singularity of which computes the 'simultaneous co-occurrences' of all quantum and relativistic (exhaustive) 'probe-target' and 'observer-phenomenon' interactions. The key revision brought about by the UCP (higher-ordered) 'A-Causal Computation' is that it negates, i.e., in principle, the existence of any 'materialistic-reductionistic SROCS "causal" relationships in QM or Relativity Theory; this is because once we accept the UCP's (higher-ordered) 'A-Causal Computation' as the sole and singular source for producing- sustaining- and evolving- any of the four phenomenal 'physical' features (of 'space', 'time', 'energy' and 'mass'), e.g., across all exhaustive spatial-pixels (in the universe) comprising any single or multiple USCF frame/s (at the minimal USCF time-point), then we must reject any 'materialistic-reductionistic SROCS' physical relationship/s between any hypothetical quantum 'probe' and 'target' or between any relativistic 'observer' and 'phenomenon' entities as "causing" any hypothetical change or effect in that subatomic target (e.g., such as the assumed "collapse of the target's probability wave function" as "caused" by its interaction with the subatomic probe element) or as "causing" any effect in the given relativistic 'phenomenon' (e.g., such as in Relativity's assumed "curvature of space-time" as "caused" by its interaction with 'massive objects'). Thus, the necessary revision in both QM and RT brought about by the UCP's 'A-Causal Computation' is to base all 'phenomenal' (quantum or relativistic) physical features of 'space', 'time', 'energy' and 'mass' on the singularity of the UCP which solely produces- sustains- or evolves- all of these physical features at all spatial-pixels in the universe comprising any (single or multiple) exhaustive USCF's frame/s...

7. The CUFT's embedding & transcendence of QM and RT models

It should, nevertheless, be made clear that this necessary revision of both Quantum Mechanical and Relativistic Models – does not negate any of the validated empirical phenomena or known quantum or relativistic laws and relationships, but rather broadens our theoretical understanding of these quantum and relativistic phenomena, as embedded in- and (indeed) transcended by- the CUFT theoretical framework; This is due to the fact that whereas Relativity Theory may represent the characterization of single spatial-temporal (relativistic) objects and phenomena, and Quantum Mechanics represents multi spatial-temporal 'probability wave function' (subatomic) entities (e.g., which also embeds 'single' multi spatial-temporal 'particle' elements) – the CUFT expands the theoretical framework to include all single- multiple- and indeed *exhaustive*- spatial-pixels comprising any minimal time-point USCF frame/s... By doing so, and based on the CUFT's identification of this minimal time-point (extremely rapid-series: 'c^2/h') series of USCF's produced solely by the singular 'Universal Computational Principle', the CUFT is capable of fully integrating between quantum and relativistic components and phenomena – which is made most apparent in the CUFT's Universal Computational Formula (e.g. that fully integrates between quantum and relativistic relationships, as well as between the four basic 'physical' features of 'space', 'time', 'energy' and 'mass'):

$$\frac{`c^2}{h} = \frac{s}{t} \times \frac{e}{m}$$

This embedding- and transcendence- of both quantum and relativistic phenomena and relationships within the broader (higher-ordered) CUFT is made most apparent in the two (above mentioned) Relativistic and Quantum formats, which include the known relativistic 'energy and mass equivalence' ($E = Mc^2$) and the quantum 'complimentary pairs';

I. Relativistic Format: $e \times \frac{s}{t} = m \times \frac{c^2}{h}$

II. Quantum Format: $t \times m \times \frac{c^2}{h} = s \times e$

Note, however, that in both quantum and relativistic formats the Universal Computational Formula transcends these (known) quantum and relativistic relationships based on the incorporation of these known relationships within the broader (quantum and relativistic) formats computational structure: Specifically, it becomes apparent that these known quantum and relativistic relationships may represent "special cases" – within the broader Relativistic or Quantum Formats, which in fact represent the complete integration of both "quantum" and "relativistic" computational components within the singular higher-ordered (fully integrated) CUFT's Universal Computational Formula; *Indeed, a more comprehensive mathematical (and empirical) validation of these two Quantum and Relativistic Formats – as different from the (above-mentioned) known quantum and relativistic relationships constituted one of the three (abovementioned) "differential-critical predictions" differentiating the CUFT from both QM and RT* [2], *and should be further investigated and validated – i.e., both mathematically and empirically.*

On the principle theoretical level, it can be pointed out that whereas the current 'Quantum Mechanical' and 'Relativistic' models represent particular phenomena and relationships – corresponding to 'single' spatial-temporal relativistic objects (and phenomena) or subatomic 'multi spatial-temporal' "wave" (and embedded 'single spatial-temporal' "particle") elements and phenomena, the CUFT fully integrates these apparently distinct, particular phenomena within the higher-ordered series of exhaustive Universal Simultaneous Computational Frames (USCF's) which is produced by the Universal Computational Principle based on its singular 'A-Causal Computation' – which produces- sustains- and evolves- all four (phenomenal) 'physical' features of 'space', 'time', 'energy' and 'mass'; Indeed, as we've seen, this singular (higher-ordered) UCP 'A-Causal Computation' negates any "materialistic-reductionistic" (SROCS) quantum or relativistic physical relationships (e.g., such as RT's SROCS assumption regarding the 'curvature of space-time' as "caused" by 'mass', or as QM's SROCS assumed 'collapse of the target's probability wave function' as "caused" by its direct physical interaction with another 'subatomic probe element'); Instead, the singularity of this higher-ordered UCP brings about the complete integration- and indeed transcendence- of the four (quantum and relativistic) phenomenal 'physical' features of 'space', 'time', 'energy' and 'mass'; A such, the CUFT goes beyond RT's integration of 'space-time' and 'energy-mass', and its curvature of space-time by mass (and vice versa) – by fully integrating 'space', 'time', 'energy' and 'mass' as four secondary computational (phenomenal) 'physical' features produced- sustained- and evolved- by the singular (computationally invariant) UCP... Likewise, the CUFT goes beyond QM's complimentary pairs of 'space and energy', 'time and mass' as constrained by Planck's constant ('h' simultaneous accuracy measurement) – as representing the exhaustive (complimentary) computational levels of the two UCP's Computational Dimensions of (Framework and Consistency) (as explained in [1]. Ultimately, the CUFT completely integrates the apparently "distinct" aspects of Quantum and Relativistic models as comprising integral computational aspects of the same singular higher-ordered UCP 'A-Causal Computation' – i.e., such as the complete integration of Relativity's 'c^2' (associated with the speed of light constraint imposed on the transmission of any signal) with Quantum's Planck's constant ('h', associated with subatomic complimentary pairs' simultaneous measurement accuracy constraint) to signify the CUFT's identified rate of UCP rapid production of the series of USCF's frames. Finally, as noted above, the CUFT's unique recognition of the singularity of the UCP's (e.g., computationally invariant) higher-ordered 'A-Causal Computation' which solely produces- sustains- and evolves- all four (e.g., computationally variant) phenomenal 'physical' features of 'space', 'time', 'energy' and 'mass' negates the current basic "materialistic-reductionistic" (SROCS) assumption underlying QM and RT – and forces Physics to recognize this singular UCP 'A-Causal Computation' as the sole reality giving rise to the phenomenology of all 'physical' features, including all quantum and relativistic phenomena and relationships.

8. The CUFT revision of 'dark energy/matter' & 'second law of thermodynamics'

One of the initial theoretical implications of the acceptance of the CUFT as a satisfactory TOE and the acceptance of its singular (higher-ordered) UCP 'A-Causal Computation' as revising contemporary Physics' Quantum and Relativistic "materialistic-reductionistic" SROCS assumption – is its capacity to explain the unresolved "enigma" of 'Dark Matter' and 'Dark Energy' and its potential revision of the 'Second Law of Thermodynamics' (and its associated 'Arrow of Time' enigma); Essentially, the 'Dark Matter, Dark Energy' enigma constitutes the inability of contemporary Physics to account for the acceleration in the rate of expansion of the physical universe – solely based on the observed (and calculated) total mass and energy associated with all planetary object comprising this physical universe... According to these calculations roughly 70-90% of all the mass and energy in the universe is "missing", i.e., cannot be observed! Hence, the working assumption (of Contemporary Physics) is that this (70-90%) of the "missing" mass and energy in the universe is "dark", i.e., it cannot be observed empirically (for some unexplained reason)...

Interestingly, this 'Dark Matter, Dark Energy' enigma is closely connected with the above mentioned UCP 'A-Causal Computation' constraining Relativity's SROCS (interactive) determination of 'massive objects' "causing" the 'curvature of space-time' and vice versa: 'curved space-time' "causing" these 'massive objects' to travel along curved space-time pathways... As delineated above, both the CUFT's 'Duality Principle' theoretical postulate and the discovery of the UCP's singular 'A-Causal Computation' prove (unequivocally) the impossibility of any such Relativity's SROCS "materialistic-reductionistic" assumption: e.g., due to such Relativistic SROCS inevitably leading to both 'logical inconsistency' and ensuing 'computational indeterminacy' which are contradicted by Relativistic Systems empirical capacity to determine both the curvature of space-time and the movement pathways of massive objects; as well as due to the UCP's 'A-Causal Computation' "minimal time-point" USCF's analysis which indicates that based on the 'Computational Invariance Principle' proof that only the 'computationally invariant' UCP exists constantly both "during" the USCF frames and also "in-between" USCF frames, whereas the four phenomenal 'physical' features of 'space', 'time', 'energy' and 'mass' only exist "during" the USCF frames as produced by the singular UCP but cease to exist "in-between" these USCF frames – we must conclude that it is not possible for any of these four phenomenal 'physical' features to "cause" any change across USCF's frames... In other words, the discovery of the UCP singular 'A-Causal Computation' (alongside the Duality Principle) negates the basic "materialistic-reductionistic" SROCS (Relativistic) current assumption, wherein it may be possible for any direct (phenomenal) 'physical' interaction, e.g., of any of these four phenomenal 'physical' features (of 'space', 'time', 'energy' and 'mass') to "cause" any change (or effect) upon another physical attribute (e.g., such as the abovementioned 'curvature of space-time by massive objects' or its vice versa: 'curved space-time' "causing" 'massive object to move along these curved space-time pathways...) Instead, the UCP asserts that the only singular 'A-Causal Computation' solely responsible for the production- sustenance- and development- of any of these four 'phenomenal physical' features (e.g., at any hypothetical 'spatial pixels' comprising any single or multiple USCF's) is singularly conducted by the UCP's computation of the "simultaneous co-occurrence" of all spatial pixels comprising any such USCF.

Hence, the UCP's singular (proven) computation of the 'simultaneous co-occurrence' of all (four) phenomenal 'physical' features of 'space', 'time', 'energy' and 'mass' comprising all spatial pixels in the universe at any minimal time-point (single or multiple) USCF's necessarily negates the possibility of any 'materialistic-reductionistic' relationship existing between any of these four (secondary computational) phenomenal 'physical' features – i.e., including both the curvature of space time by massive objects (or vice versa) as well as the "expansion of the physical universe" – as "caused" by the phenomenal features of the amount of "mass" or "energy" comprising any single or multiple 'USCF' frame/s... In other words, based on the CUFT's proven singularity of the UCP – in producing- sustaining- and developing- all (four) phenomenal 'physical' features of 'space', 'time', energy' and 'mass' (across all exhaustive spatial pixels comprising the totality of the physical universe at any minimal time-point single or multiple USCF frame/s), the UCP's 'A-Causal Computation' is seen as solely responsible for all quantum, relativistic and CUFT known (or predicted) phenomena: This includes also the observed accelerated expansion of the physical universe – i.e., which indeed cannot be accounted for through any 'materialistic-reductionistic ' interactions between any of these secondary computational phenomenal 'physical' features. Indeed, viewed from this singular (higher-ordered) perspective of the UCP'S sole production- sustenance- and evolution- of all spatial pixels in the universe (comprising any minimal time-point single or multiple USCF frame/s) all quantum and relativistic phenomena, e.g., including the accelerated expansion of the physical universe must be accounted for solely through the UCP Causal Computation; Hence, according to the CUFT the relativistic phenomenon of the accelerated expansion of the physical universe cannot be accounted for by the currently assumed "Dark Energy and Dar Matter" – which represent a "materialistic-reductionistic" assumption (as explained in detail above), but instead must be explained as arising from the sole and singular production- sustenance- and evolution- of the physical universe by the UCP...

Another interesting potential theoretical ramification of the adoption of the CUFT as a satisfactory 'TOE' – including its discovery of the UCP singular 'A-Causal Computation', may be its potential revision of the 'Second Law of Thermodynamic' (and associated 'Arrow of Time' phenomenon). The 'Arrow of Time' enigma refers to the observation that the laws of Physics are "biased" in such a manner that events (and phenomena) always occur in a uni-directional temporal direction: thus, for instance, a glass may break into a hundred pieces – but those hundred pieces will not (of themselves) revert back to form a single unitary glass... Indeed, closely associated with this 'Arrow of Time' unidirectional temporal characteristic of physical phenomena is the (famous) 'Second Law of Thermodynamics' which states that in any given physical system the degree of entropy always increases with time... However, based on the CUFT's discovery of the singular UCP 'A-Causal Computation' – which was shown (above) to negate the basic "materialistic-reductionistic" assumption (underlying both Quantum and Relativistic models of physical reality), and one of the CUFT's (previous: [2] 'differential-critical' predictions regarding the possibility of "reversing the sequence of spatial-electromagnetic pixels" based on the application of certain electromagnetic effects, we may need to revise this Second Law of Thermodynamics (and associated 'Arrow of Time' enigma); This is because as explained earlier, none of the quantum or relativistic physical phenomena, relationships (or even laws) can continue to be based on any "materialistic-reductionistic" assumption/s. Hence, as shown above, neither the curvature of space-time by massive objects or (vice versa) the determination of the movement of massive objects based on the curvature of space-time, nor the observed accelerated expansion of the physical universe – can be explained by the current 'materialistic-reductionistic' relativistic (or quantum) assumption, but must be based on the singularity of the UCP 'A-Causal Computation'; Indeed, as we've shown (above), a fine temporal analysis of the dynamics of this singular UCP's production-sustenance- and evolution- of every (exhaustive) spatial pixel in the universe (comprising any single or multiple USCF frame/s) indicates that there cannot exist any 'materialistic-reductionistic' effect of any of the four phenomenal 'physical' features (of 'space', 'time', 'energy' or 'mass') between any two (or more) spatial pixels, i.e., either within the same USCF frames or across different USCF frames. This is due to the UCP's singular asserted computation of the "simultaneous co-occurrence" of all spatial pixels in a given USCF frame (which prohibits any "causal" materialistic-reductionistic" effects existing between any two or more spatial pixels in the same USCF frame), as well as the UCP 's 'A-Causal Computation' associated 'Computational Invariance Principle' which indicates that the sole and singular reality existing invariantly "during" the USCF's frame/s (e.g., as producing, sustaining and evolving any of the four phenomenal 'physical' features of all of its spatial pixels) and "in-between" these USCF's frames is the UCP. Hence, the only source for producing- sustaining- and evolving- any spatial pixel in the physical universe (e.g., at any given USCF frame/s) is the singular UCP, but not any of its (computationally variant) phenomenally produced 'physical' features...

Therefore, also the 'Second Law of Thermodynamics' which asserts the increase in entropy of any physical system with the progression of time – must be revised based on this new higher-ordered recognition of the singularity of the UCP 'A-Causal Computation': Hence, instead of the currently assumed 'materialsitic-reductionistic' basis for this Second Law of Thermodynamics, i.e., wherein it is the physical relationships that exist between a given physical system's material components which "causes" the degree of entropy in that system to necessarily increase with time, the UCP's 'A-Causal Computation' points unequivocally at the singularity of the UCP as producing- sustaining- and evolving- all spatial pixels in the universe and all associated physical phenomena and laws... Moreover, since all four phenomenal 'physical' features of 'space', 'time', 'energy' and 'mass' – are shown to comprise only 'computationally variant' features singularly produced by this UCP then we can foresee a condition in which the spatial-temporal sequence of a given USCF's frames can be reversed (i.e., at least when it is limited to a particular physical phenomenon); This was indeed predicted as one of the CUFT's 'differential-critical' predictions [2] – i.e., regarding the possibility of reversing the 'spatial-electromgnetic' sequence of a given phenomenon such as the growth and decay of a given amoeba. Essentially, this 'critical-differential' prediction of the CUFT states that it should be possible at least in principle) to reverse any given physical phenomenon by recording its precise USCF's spatial-electromagnetic values (e.g., of each of its constituting spatial pixels across a give number of USCF's frames), and then applying a specific electromagnetic stimulation (to each of this phenomenon or physical object's spatial pixels) in such a manner as to produce the "reversed spatial electromagnetic sequence" across the same number of given USCF's frames! Therefore, this CUFT's 'differential-critical' prediction predicted that it should be possible (at least in principle) to "cause" an 'amoeba' to "go back in time" – reversing its spatial-electromagnetic spatial pixels' sequence (by applying the particular electromagnetic stimulation to each of its spatial-pixels across a given number of USCF's frames... More generally then, the CUFT asserts the possibility of reversing the sequence of temporal events comprising any physical phenomenon! Therefore, it should be possible to increase the degree of entropy in any given physical system – contrary to the (currently accepted) 'Second Law of Thermodynamics'!

Although apparently "radical" this 'differential-critical' prediction of the CUFT does not aim to "topple down" the foundations of theoretical Physics, but rather expand our understanding of the physical reality by incorporating both Quantum Mechanics and Relativity Theory within a broader (higher-ordered) theoretical framework based on the discovery (and initial empirical verification) of the CUFT and its associated singularity of the UCP 'A-Causal Computation'; This is simply because in light of contemporary Physics basic contradiction between its two primary theoretical pillars (e.g., Quantum Mechanics and Relativity Theory) which has been shown to be resolved by the CUFT, the (initial) empirical validation of the CUFT's 'differential-critical' prediction associated with the 'Proton-Radius Puzzle', and its discovery of the singularity of the UCP's 'A-Causal Computation' – the fundamental concepts of 'space', 'time', 'energy' and 'mass' as representing merely secondary ('computationally variant') 'phenomenal' features produced by the sole reality of the ('computationally invariant') UCP have to be revised: Specifically, since "time" (alongside all three other 'phenomenal' physical features) is conceptualized as being singularly produced by the UCP – e.g., representing the degree of change of any given object or phenomena across a series of USCF's frames, then it should be possible (at least in principle) to reverse the sequence of spatial-change across frames (through the application of specific electromagnetic stimulation to the relevant spatial pixels comprising this physical phenomenon), thereby reversing the temporal events comprising this physical phenomenon... Hence, it may be said that the Second Law of Thermodynamics accurately represents the "natural progression" or temporal phenomena – but must be revised to include the possibility of reversing these 'natural phenomena' (thereby increasing their measured degree of entropy) across a series of USCF's frames. In a broader theoretical sense, the discovery of the CUFT's 'A-Causal Computation' necessitates us to revise our basic 'materialistic-reductionistic' assumptions underlying contemporary Physics, in such a manner that Quantum Mechanics and Relativity Theory will be anchored and based on the singular higher-ordered operation of the UCP's A-Causal Computation...

Therefore, we see that there is an urgent need to revise both quantum and relativistic models (laws and phenomena) based on the CUFT's discovery of the singularity of the UCP's production- sustenance- and evolution- of the physical universe; This important task involves several future steps, including: an empirical and mathematical verification of all of the CUFT's "differential-critical" predictions (e.g., beyond the initial empirical validation of one of its 'differential-critical' predictions associated with the 'Proton-Radius Puzzle' findings, mentioned earlier), a revision of the laws of Physics based on the CUFT's 'Universal Computational Formula' (which in fact fully embeds and integrates the key quantum and relativistic components) and further explication and exploration of the new theoretical vistas offered by the CUFT higher-ordered and broader theoretical framework (including the potential connection between this singular Universal Computational/Consciousness Principle and individual human Consciousness).

Author details

Jehonathan Bentwich

Address all correspondence to: drbentwich@gmail.com

'BLIS' LTD, Israel

References

[1] Bentwich, J. (2012a) "Harmonizing Quantum Mechanics and Relativity Theory". *Theoretical Concepts of Quantum Mechanics*, Intech (ISBN 979-953-307-377-3), Chapter 22, pp. 515-550.

[2] Bentwich, J. (2012b) "Theoretical Validation of the Computational Unified Field Theory". Theoretical Concepts of Quantum Mechanics, (ISBN 979-953-307-3773), Chapter 23, pp. 551-598.

[3] Bentwich, J. (2013a). "The Theoretical Ramifications of the Computational Unified Field Theory". Advances in Quantum Mechanics (ISBN 978-953-51-1089-7), Chapter 28, pp. 671-882.

[4] Bentwich, J. (2013b). The Computational Unified Field Theory (CUFT): A Candidate Theory of Everything. Advances in Quantum Mechanics (ISBN 978-953-51-1089-7) Chapter 18, pp. 395-436.

[5] Bentwich, J. (2014a) What if Einstein was Right? *Amazon Kindle Book Store.*

[6] Bentwich, J. (2014b) The Next Scientific Shift: The Computational Unified Field Theory. Intech Publication (In Press).

[7] Bernauer & Pohl (2014). The Proton Radius Puzzle. *Scientific American,* 310 (2), p. 20-24.

[8] Bentwich, J. The 'Duality Principle': Irreducibility of sub-threshold psychophysical computation to neuronal brain activation. *Synthese,* Vol. 153, No. 3, pp. (451-455) (2006a).

[9] Bentwich, J. *Universal Consciousness: From Materialistic Science to the Mental Projection Unified Theory,* iUniverse Publication (2006).

[10] 10. Born, M. The statistical interpretation of quantum mechanics, *Nobel Lecture, December* 11, *1954.* Brumfiel, G. Our Universe: Outrageous fortune. *Nature,* Vol. 439, pp. (10-12) (2006).

[11] Ellis, J. The Superstring: Theory of Everything, or of Nothing? *Nature,* Vol. 323, No. 6089, pp. (595–598) (1986).

[12] Greene, B. *The Elegant Universe,* Vintage Books, New York (2003).

[13] Heisenberg, WÜber den anschaulichen Inhalt der quantentheoretischen Kinematik und Mechanik. *Zeitschrift für Physik,* Vol. 43 No. 3–4, pp. (172–198) (1927).

[14] D. Hilbert. 'Die Grundlagen Der Elementaren Zahlentheorie'. *Mathematische Annalen* 104:485–94. Translated by W. Ewald as 'The Grounding of Elementary Number Theory', pp. 266–273 in Mancosu (ed., 1998) From Brouwer to Hilbert: The debate on the foundations of mathematics in the 1920's, Oxford University Press. New York.

[15] Horodecki, R.; Horodecki, P.; Horodecki, M. & Horodecki, K. Quantum entanglement. *Rev. Mod. Phys,* Vol. 81, No. 2, pp. (865–942) (2007).

[16] "Ockham's razor". Encyclopædia Britannica. Encyclopædia Britannica Online. http://www.britannica.com/EBchecked/topic/424706/Ockhams-razor. Retrieved 12 June 2010.

[17] Polchinski, J. All Strung Out? *American Scientist,* January-February 2007 Volume 95, Number 1 (2007).

[18] Stephen W. Hawking Godel and the end of Physics. (Public lecture on March 8 at Texas A&M University) 2003.

[19] On the right track. Interview with Professor Edward Witten, *Frontline*, Vol. 18, No. 3, February 2001.

Permissions

All chapters in this book were first published in STAQM, by InTech Open; hereby published with permission under the Creative Commons Attribution License or equivalent. Every chapter published in this book has been scrutinized by our experts. Their significance has been extensively debated. The topics covered herein carry significant findings which will fuel the growth of the discipline. They may even be implemented as practical applications or may be referred to as a beginning point for another development.

The contributors of this book come from diverse backgrounds, making this book a truly international effort. This book will bring forth new frontiers with its revolutionizing research information and detailed analysis of the nascent developments around the world.

We would like to thank all the contributing authors for lending their expertise to make the book truly unique. They have played a crucial role in the development of this book. Without their invaluable contributions this book wouldn't have been possible. They have made vital efforts to compile up to date information on the varied aspects of this subject to make this book a valuable addition to the collection of many professionals and students.

This book was conceptualized with the vision of imparting up-to-date information and advanced data in this field. To ensure the same, a matchless editorial board was set up. Every individual on the board went through rigorous rounds of assessment to prove their worth. After which they invested a large part of their time researching and compiling the most relevant data for our readers.

The editorial board has been involved in producing this book since its inception. They have spent rigorous hours researching and exploring the diverse topics which have resulted in the successful publishing of this book. They have passed on their knowledge of decades through this book. To expedite this challenging task, the publisher supported the team at every step. A small team of assistant editors was also appointed to further simplify the editing procedure and attain best results for the readers.

Apart from the editorial board, the designing team has also invested a significant amount of their time in understanding the subject and creating the most relevant covers. They scrutinized every image to scout for the most suitable representation of the subject and create an appropriate cover for the book.

The publishing team has been an ardent support to the editorial, designing and production team. Their endless efforts to recruit the best for this project, has resulted in the accomplishment of this book. They are a veteran in the field of academics and their pool of knowledge is as vast as their experience in printing. Their expertise and guidance has proved useful at every step. Their uncompromising quality standards have made this book an exceptional effort. Their encouragement from time to time has been an inspiration for everyone.

The publisher and the editorial board hope that this book will prove to be a valuable piece of knowledge for researchers, students, practitioners and scholars across the globe.

List of Contributors

M.R. Pahlavani
Department of Nuclear Physics, Faculty of Basic Science, University of Mazandaran, Babolsar, Iran

Maricel Agop
Lasers, Atoms and Molecules Physics Laboratory, University of Science and Technology, Lille, France
Department of Physics, "Gh. Asachi"' Technical University, Iaşi, România

Alina Gavriluţ
Department of Mathematics, "Al.I. Cuza" University from Iaşi, Romania

Lăcrămioara Ochiuz
Faculty of Pharmacy, Department of Pharmaceutical Technology University of Medicine and Pharmacy "Gr.T. Popa", Iaşi, România

Dan Tesloianu
IV Internal Medicine Department, Clinical Emergency Hospital "Sf. Spiridon", University of Medicine and Pharmacy "Gr.T. Popa", Iaşi, România

Gabriel Crumpei
Psychiatry, Psychotherapy and Counselling Center, Iaşi, România

Cristina Popa
Dentistry Department, University of Medicine and Pharmacy "Gr.T. Popa", Iaşi, România

Daniel Timofte
Surgery Dept., Hospital "Sf. Spiridon", University of Medicine and Pharmacy "Gr. T. Popa", Iaşi, Romania

Lucian Eva
Emergency Clinical Hospital "Prof. Dr. Nicolae Oblu", Iaşi, Romania

Decebal Vasincu
Surgery Dept., Biophysical Section, University of Medicine and Pharmacy "Gr. T. Popa" Iaşi, Romania

Călin Gh. Buzea
National Institute of Research and Development for Technical Physics - IFT Iasi, Iaşi, Romania

Maricel Agop
Dept. of Physics "Gh. Asachi" Technical University, Iaşi, Romania
Lasers, Atoms and Molecules Physics Lab., University of Science and Technology, Lille, France

Radu Florin Popa
Surgery Dept., University of Medicine and Pharmacy "Gr. T. Popa", Iaşi, Romania

V.I. Sbitnev
St. Petersburg B.P. Konstantinov Nuclear Physics Institute, NRC Kurchatov Institute, Gatchina, Leningrad district, Russia
Department of Electrical Engineering and Computer Sciences, University of California, Berkeley, USA

Cynthia Kolb Whitney
Galilean Electrodynamics, USA

Aghaddin Mamedov
SABIC Technology Center, Sugar Land, TX, United States

Jehonathan Bentwich
'BLIS' LTD, Israel

Index

R

S

T

U

V

W

www.ingramcontent.com/pod-product-compliance
Lightning Source LLC
LaVergne TN
LVHW060511210926
848485LV00005B/1065